# Governing Food

## Science, Safety and Trade

# The Canada-United Kingdom Colloquia Series

# Governing Food

## Science, Safety and Trade

Peter W.B. Phillips and Robert Wolfe, Editors

Published for the School of Policy Studies, Queen's University
by McGill-Queen's University Press
Montreal & Kingston • London • Ithaca

**National Library of Canada Cataloguing in Publication Data**

Main entry under title:

Governing food : science, safety and trade

(The Canada-United Kingdom colloquia series)
Includes bibliographical references.
ISBN 0-88911-903-1 (bound).--ISBN 0-88911-897-3 (pbk.)

1. Food--Safety measures. 2. Food--Safety regulations.
3. Food--Government policy. 4. Food industry and trade.
I. Phillips, Peter W. B. II. Wolfe, Robert, 1950-   III. Queen's
University (Kingston, Ont.). School of Policy Studies. IV. Series:
Canada-United Kingdom colloquia series.

TX537.G68 2001        363.19'2        C2001-903004-5

# The Canada-UK Colloquia

The Canada-UK Colloquia are annual conferences that aim to increase knowledge about important issues of concern to Canada and the United Kingdom. These conferences take place alternately in each country, bringing together British and Canadian parliamentarians, public officials, academics, representatives from the private sector, graduate students, and others. One of the main endeavours is to stimulate and publish research on each subject under discussion and the publications listed on the back of the title page demonstrate the wide range of topics covered by recent Colloquia.

The Colloquia are supported by the Department of Foreign Affairs and International Trade in Canada and by the Foreign and Commonwealth Office in the United Kingdom. The conferences are organized by the School of Policy Studies at Queen's University on the Canadian side; and by the Canada-UK Colloquia Committee on the British side.

The first Colloquium, attended by some 60 participants from both countries, was held at Cumberland Lodge in Windsor Great Park in 1971 to examine the bilateral relationship. This theme figured in the Colloquium held at Leeds University in 1979, at Dalhousie University in 1984, and again at Queen's University in 1996. A British steering committee, later to become the British Committee, was launched in 1986. The School of Policy Studies assumed responsibility on the Canadian side in 1996.

At the Denver Summit in June 1997, Prime Ministers Blair and Chrétien issued a Joint Declaration to mark a program of modernization in the bilateral relationship which included a role for the Canada-United Kingdom Colloquia. The program was reaffirmed during Mr. Chrétien's visit to the UK in 1998.

More information is available at the Colloquium Web site: www.canada-uk.net.

# Contents

## SAFETY: RISK COMMUNICATION

## CONCLUSION

# Foreword

Among consumer products, food is unique. It is unique because of its very nature as being essential to life. It is unique because of the fact that it is a very heterogeneous commodity. It comes in all shapes and sizes; it may be a direct product of farming or fishing; it may be a mixture of many ingredients or raw materials made in small quantities or from large-scale manufacture; it may be the result of knowledge of food technology or advances in the scientific knowledge of breeding animals or growing crops. Farming and fishing are ultimately directed at food production and scientific developments in the use of chemicals or new methods in animal or crop production can impact directly on the nature and substance of the food produced and supplied for human consumption. The consumer has an awareness of all these matters, albeit often fairly rudimentary, and that in its turn contributes to the uniqueness of food as a consumer product. Consumers have a different approach to food compared to other consumer products. Whilst there are safety risks associated with items such as electrical goods or motor cars, food safety is regarded more intensely than other consumer product risks. Why this is so is not absolutely clear; perhaps it has something to do with the ingestion of food. Other products remain external in their risks to health or safety. Food is internal. In their approaches to food safety policies, governments have to recognize this special place that food occupies for consumers.

The need for the governance of food is not new. Primitive societies recognized early on that it was wise to develop and lay down rules relating to the production and preparation of food. The *Book of Leviticus* contains a whole series of food hygiene and safety rules whose aim is clearly health protection amongst the early Israelites. This is evidence that authority does need to set down rules for the protection of individuals from those in food production or handling who would expose people to risk by careless unhygienic practices or would deceive people through adulteration or misdescription. The earliest known food law in the United Kingdom is in this latter category — the *Assize of Bread and Ale* of 1265 was to prohibit the use of chalk instead of flour in bread and the watering down of beer.

The need for complex rules was less pressing during the many centuries that mankind remained essentially a rural population with people living off their home or local produce. But with the Industrial Revolution and the growth of large urban populations consumers became divorced from producers by long chains of supply and large-scale food manufacturing and handling. Governments were in those circumstances required to devise, promulgate, and enforce food law structures which would protect the health and welfare of their populations. Thus, the nineteenth century saw the development of the current base of food law which operates in most developed countries in Europe, North America, and elsewhere in the world. Another factor that came into play at this time was the rapid development of science and technology and in the food industry this is now very significant. It is used in large-scale manufacturing to provide a plentiful food supply at affordable prices. But it brings with it more complexity and thus a need for more complex rules to protect the citizen. Governments have to bring into their policy-making a knowledge of food science and food technology together with advice on various health matters, nutrition, and toxicology in order to ensure that rules are current and relevant. They also have to bring into the equation the need to explain complex science to laypeople and thereby to obtain and retain the confidence of consumers (who are also voters) that law-making is safeguarding their interests.

Food policies need to have a firm base if they are to be consistent and if they are going to have the necessary capacity to develop in order to meet various changes that come about in society and science. The most significant change in recent years has been the shift from a concern with producing a sufficient food supply to worries about healthy eating and what are often called "manufactured foods." As a result government food policies have become less production-orientated and more consumer-orientated. This can be seen in both Europe and North America where policy responsibilities have moved from agricultural directorates to health or consumer protection directorates. But the problem is that to adopt a policy base of consumer preference is far less certain than one based on food science. Science may well not be complete and it may in time change, but consumer preference is much more fickle and diverse and a policy based on it can find itself regularly exploring blind alleys. Thus, a balance has to be struck between the relative certainty of science (even with the uncertainties that new scientific developments may contain) and consumer confidence that it is not at risk from issues it does not fully understand.

So what does this mean in overall terms of the governance of food? The governance of food is not solely a matter for governments. All individuals and companies involved in the production, manufacture, distribution, and marketing of food have a part to play. Governments clearly have the responsibility to set up frameworks that contain not only the legal rules but also systems for the testing and assessment of foodstuffs and ingredients — especially those that are new — to check that such products can safely be consumed or where such approval has to be conditional that those conditions are known and observed. Governments must therefore also have enforcement and inspection systems. But no official system can provide 100 percent coverage. Operators have the first responsibility to ensure that the safety rules are carefully followed, that resources are put into production lines to see that this is done and that consumers are properly informed about the nature of the product on offer. In the interest of continuing to trade, most operators do, of course, have a self-interest to do all of this.

Thus the real issue currently surrounding the governance of food is one of consumer confidence in the food supply. It is one of establishing a rapport between producer and consumer; of explaining what are the benefits that the consumer gets from many of the scientific procedures such as longer shelf life to allow less frequent shopping or out-of-season produce or of convenience in preparation and cooking times. Unless this rapport can be re-established it does seem possible that governments will retreat into an over-regulatory mode which could frustrate scientific developments to the ultimate advantage of no one. So the question for everyone is how does every player in the food chain help to restore confidence in the modern food supply whilst facilitating scientific development? How do farmers, manufacturers, retailers, politicians and administrators convince the customer that his or her interest is supreme and nothing is being done that will jeopardize that? And last, but not least, how can consumers educate themselves to better understand what is happening and thus discern the true voices from the false?

*Charles Cockbill*
*President*
*European Food Law Association*

# Preface

Trade in food is one of the oldest aspects of the close relations between Canada and the United Kingdom. In this era of global food markets, however, UK imports of Canadian beef are negligible, and Canada refuses to import beef from the UK. Food safety is the reason in both cases. The EU ban on imports of beef from cattle treated with growth hormones has devastated Canadian markets in the UK, and the lingering problems with mad cow disease have devastated British beef exports to the world. It is not surprising, therefore, that food safety and the governance of the food system represented a compelling topic for the 2000 Canada-United Kingdom Colloquium. Both Canada and the United Kingdom confront similar policy challenges rooted in domestic concerns and pressures on the one hand and the global nature of the food system on the other. However, the two countries also bring distinctive views to this debate, reflecting different cultural assumptions, different recent experiences with problems of food safety, and memberships in different regional organizations, the European Union and NAFTA.

The 2000 Colloquium provided an opportunity for political leaders, senior officials, business people, academics, and representatives of civil society organizations from both countries to consider this extraordinarily complex issue. Saskatoon was a wonderful location for their discussions. The city is one of the leading centres of agricultural research in the world, and home to many companies at the cutting edge of agricultural change. In addition, the province of Saskatchewan is a major producer of food for global markets.

The Colloquium was possible only because of the continued support of our core sponsors, the Department of Foreign Affairs and International Trade in Canada, and the Foreign and Commonwealth Office in the United Kingdom. In addition, the 2000 Colloquium was supported by the University of Saskatchewan, the College of Agriculture, the Agriculture Biotechnology Initiative, and the Virtual College of Biotechnology. We are deeply grateful for their assistance.

We wish to record our appreciation of the contribution of the Honourable Allan Blakeney, who chaired the proceedings superbly through two days of the

Colloquium, and of Professor William Leiss, who served as rapporteur and subsequently prepared a report.

Special thanks are also due to this year's organizers, led on the Canadian side by Professors Robert Wolfe at Queen's University and Peter Phillips and George Khachatourians at the University of Saskatchewan. They, in turn, are grateful for the advice they received from Professors William Leiss and Douglas Powell. On the British side, Mr. Charles Cockbill advised the British Committee and Peter Newton, Executive Secretary of the Committee, maintained liaison with the Canadian side. We were all dependent on the detailed logistical work of Todd Yates at Queen's University. The quality of our deliberations depended on the effectiveness of the work of all these people.

*Keith Banting*  
*Director*  
*School of Policy Studies*

*Baroness Fookes of Plymouth*  
*Chairman, British Committee*  
*Canada-United Kingdom Colloquia*

# Acknowledgements

The books published in the Canada-UK Colloquium series are all subject to a review process. We are grateful to our colleagues who willingly and expertly assisted us in this process: Sir Nicholas Bayne, Jill Hobbs, William Kerr, Bruno Larue, Gregory Marchildon, Elizabeth Moore, and Grace Skogstad. We are especially grateful to Danielle Koster for her invaluable assistance in editing the texts, and to Alex van Kralingen who found new public opinion data for us.

The professionalism of the staff of the Publications Unit of the School of Policy Studies made our task as editors much easier. Marilyn Banting carried out the copyediting, Valerie Jarus set the type, and Mark Howes managed the process and designed the cover.

*Peter W.B. Phillips*
*Robert Wolfe*

# Acronyms

| | |
|---|---|
| AIA | Advance informed agreement |
| AWSA | Agrichemical Warehousing Standards Association |
| BEUC | European Organization of Consumers Unions |
| BGH | Bovine growth hormone |
| BSE | Bovine spongiform encephalopathy |
| BSP | BioSafety Protocol |
| BSWG | Biosafety Working Group |
| Bt | Bacillus thuringienis |
| CBD | Convention on Biological Diversity |
| CCFL | Codex Committee on Food Labelling |
| CFIA | Canadian Food Inspection Agency |
| CGSB | Canadian General Standards Board |
| Codex | *Codex Alimentarius* Commission |
| CPI | Crop Protection Institute |
| DNA | Deoxyribonucleic Acid |
| ECB | European corn borer |
| EFA | European Food Authority |
| EU | European Union |
| FAO | Food and Agriculture Organization |
| FSA | Food Standards Agency, UK |
| GATT | General Agreement on Tariffs and Trade |
| GE | Genetically engineered |
| GM | Genetically modified |
| GMO | Genetically modified organism |
| GMP | Good Manufacturing Practices |
| GPS/GIS | Global Positioning System/Global Information System |

| | |
|---|---|
| HACCP | Hazard Analysis Critical Control Points |
| ICCP | Intergovernmental Committee for the Cartagena Protocol |
| ILO | Intensified livestock operations |
| IPCC | Intergovernmental Panel on Climate Change |
| IPM | Integrated Pest Management |
| IPPC | International Plant Protection Convention |
| ISO | International Organization for Standardization |
| ISPMs | International Standard for Phytosanitary Measures |
| LMOs | Living modified organisms |
| MOU | Memorandum of Understanding |
| MRL | Maximum Residue Limit database |
| NGO | Non-governmental organization |
| OECD | Organisation for Economic Co-operation and Development |
| OIE | Office international des épizooties |
| PCR | Polymerase Chain Reaction |
| R&D | Research and Development |
| rbST | recombinant bovine somatotropin |
| SPS | Sanitary and Phytosanitary Agreement |
| TBT | Technical Barriers to Trade Agreement |
| TEP | Trans-Atlantic Economic Partnership |
| WHO | World Health Organization |
| UHT | Ultra high temperature treatment of milk |
| UNCED | United Nations Conference on Environment and Development |
| UNEP | United Nations Environment Program |
| vCJD | new variant Creutzfeldt-Jakob disease |

# 1

# Governing Food in the 21$^{st}$ Century: The Globalization of Risk Analysis

*Peter W.B. Phillips and Robert Wolfe*

Food is both simple — it is what we eat to sustain life — and extraordinarily complex — what we eat includes whole and processed plants, animals and fish from all over the world, produced in a bewildering variety of ways. Food presents unique governance challenges. Assuring the safety, quality, and environmental sustainability of the food supply is both a core responsibility of government, and something that requires the active engagement of everyone in the long supply chain from farm to fork. Food safety, moreover, is the latest flash point on the global agenda. It irritates transatlantic relations (e.g., beef hormones and antibiotics), Britain's relations with its European Union (EU) partners (e.g., bovine spongiform encephalopathy [BSE]), and Canada's relations with the United Sates (e.g., food inspection systems for meat and wheat).

Food governance is complicated by the fact that it is both a globalized commodity and a fundamental part of local human societies. Food is therefore emblematic of many contemporary policy issues, where some aspect of what is conventionally called globalization — changes in technology, transportation, and communications — challenges the social norms in our communities. The public controversies that follow the globalization of production, distribution, and regulation in this industry are at the core of this book.

The politics of governing food at first glance seems to be driven by protectionism, mistrust of modern science, lack of respect for experts, inability to understand and manage risk, dislike of big business, lack of confidence in government, or by garden variety "globaphobia." But the political turmoil is also driven by the reality that millions of people in North America and Europe suffer food-related illnesses every year. The US Centre for Disease Control estimates that tens of millions of people in the United States alone are afflicted each year, and that thousands die of food-borne pathogens. Some of the pathogens are old and some are new while many are compounded by allergic reactions. These changes are related to where food comes from, how it is grown, the way it is processed, its packaging and marketing, and the means of transporting it to the table.

The authors of the chapters in this book have considered the challenges for governance in the domain of food safety from a variety of perspectives. These challenges involve science, safety, and trade. While these three domains are not mutually exclusive, they offer different ways of understanding the problems of food. Accordingly we have used the triangular tension between these ideas as the conceptual basis for organizing the book, relating each point of the triangle to one of the three aspects of the consensual understanding of *risk analysis*. In Part 1, George Khachatourians discusses science and the implications of changing scientific understanding for *risk assessment*. The regulatory and *risk management* implications of changing commercial practices are the focus in Part 2, with contributions by Neville Craddock, Robert Falkner, Lorne Hepworth, Anne MacKenzie, and Peter Phillips. Finally, in Part 3, Spencer Henson, Catherine Humphries, Patricia Mann, and Douglas Powell with Katija Blaine, Amber Leudtke, Shane Morris and Jeff Wilson assess how changing public perceptions of risk assessment and management alter the context for *risk communications*. In the conclusion, William Leiss first discusses the contrast between "science-based regulation" and "consumer sovereignty"; and then the tension between "risk assessment" and "other factors in decision-making."

Our conceptualization of the food safety system is illustrated in Figure 1. There we see that the tension between science, safety, and trade is replicated in a number of other ways of looking at the problem of food safety. The three points of the triangle also correspond to the tension between experts, individuals (citizens and consumers), and producers. Similarly we see different modes of reasoning deployed at the poles. Risk assessment uses science-based probabilistic reasoning where risk communications uses safety-oriented deterministic reasoning and risk management responds primarily to market reasoning. Perhaps most important for our theme, borders are least relevant at the science pole where members of the "community of science" tend to have shared views on risk assessment no matter where they live. Divergences are greater among governments on risk management, and place matters most for local communities, where we find that differences in the domain of risk communication can be a significant source of conflict. None of these perspectives tells the whole story on its own, and no neat line divides them, but these different frames of reference do help in thinking about the complex tensions in governing food.

We can visualize each of these frames as triangles of different and increasing size. Globalization — an expansion in the role of the market — extends all the

FIGURE 1: Elements of the Food Safety System

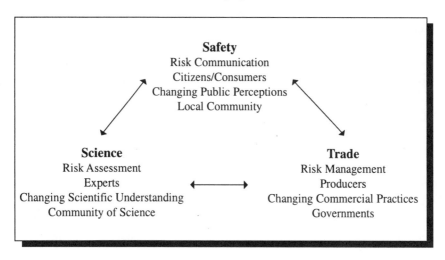

lines of the triangle, enlarging the domain of potential interaction and conflict. Globalization also creates multiple permutations and combinations of the framing factors. For example, when we consider the key actors, the tension between experts, producers, and citizens differs from that between experts, producers, and consumers. While governance requires individuals to act collaboratively as "citizens" and not independently as "consumers," what citizens say they want at home and the revealed preferences of consumers in the global market are not always the same. This divergence between our individual roles as consumers and citizens is at the root of many of the local and global disputes about food safety.

Similarly, the points of the triangle can change in importance in different countries over time. While expertise is increasingly questioned everywhere, Europeans are especially sceptical. The outbreak of bovine spongiform encephalopathy (BSE) or "mad cow disease" and the related rise of new variant Creutzfeldt-Jakob disease (vCJD) explains in part their scepticism. It has also contributed to a more cautious response than in North America to the development of biotechnology and the introduction of genetically modified organisms (GMOs). The challenge of distinguishing simple protectionism from good-faith regulation in the public interest is also seen clearly in the (in)famous beef hormones case. In brief, in the early 1980s North American regulators approved a number of hormones that could be used to promote growth in cattle, allowing farmers to reduce their feed costs and the time required to fatten animals for market. The same hormones were banned in Europe, in part because of consumer concerns about safety. As long as these contrary decisions only had local market implications, other countries were not concerned. But one consequence of the EU ban was that beef from cattle treated with the hormones could not be imported into Europe, forcing North American producers to divert exports to other markets, which increased competition for many other meat-producing countries. The resulting trade dispute is still not resolved after two decades. It has involved lengthy debates about the underlying science in technical international organizations, as the two sides have hurled opposing scientific studies at each other. This bitter trade dispute has stymied trade negotiators in the old General Agreement on Tariffs and Trade (GATT) system and now in the World Trade Organization (WTO).

Government action on hormones and GMOs has been driven by different actors and different perspectives in the two markets. In North America, producer interests largely drove governments. In Canada, to the extent anybody noticed, the hormones story was about efficient production; in Europe, it was a story about food safety and consumer sovereignty. What is especially interesting, however, is the way in which the food safety concerns of consumers and the protectionist interests of producers can and do interact. Small producers in many countries, for example, do not like to see hormones used in beef or milk production because that increases their disadvantage with respect to larger and more efficient producers. Furthermore, some producers could not use the new technology because they relied on range feeding rather than rationed feed. Argentina, for example, supported the hormones ban because all of the beef they export to Europe is grass fed.

The two systems also differ in how they ensure accountability in the regulatory process. In North America, we rely on independent administrative agencies staffed by scientists to make decisions on the basis of "sound science." The scientists are said to be objective, arm's length, and transparent. In Europe, by contrast, politicians are ultimately accountable for regulatory decisions and they are tending to rely on the "precautionary principle" when faced with uncertainty. They justify their cautious approach by noting that experts are divided among themselves, sometimes are disastrously wrong (past assurances about BSE are often cited), and often fail to address the concerns of citizens. Although independent scientists are increasingly consulted in Europe, notably in the new UK Food Safety Agency, they lack the power held by the scientists at the Canadian Food Inspection Agency or at the US Food and Drug Administration.

The interaction of science, trade, and safety and the related connections between different actors in different domains, largely defines the current set of issues facing those attempting to govern food in the twenty-first century. In the rest of this chapter, and the rest of the book, we examine the elements of each point of our conceptual triangle.

## SCIENCE

At one level, the theme for this book could be expressed as: What do policymakers need to know about the science of food safety and quality, and about new food technologies? It could equally be expressed as: What do scientists need to know about policy? In short, how do scientific controversies affect national and global governance? Answers to these questions are important for the first part of risk analysis, risk assessment, which the *Codex Alimentarius* (Codex) defined as "a scientifically-based process consisting of the following steps: (i) hazard identification, (ii) hazard characterization, (iii) exposure assessment, and (iv) risk characterization."[1] But what is "a scientifically based process?" The risk assessment process engages academic and government experts who collaborate in the generation of a consensual understanding of the issues with testing organizations, national expert panels, Food and Agriculture Organization (FAO)/World Health Organization (WHO) Expert Committees, international scientific bodies such as the International Plant Protection Convention and international organizations like the Organisation for Economic Co-operation and Development (OECD). While this process depends on open dissemination of scientific information, it becomes more complex as expertise fragments and specializes, and as new products create new hazards.

The appeal to "science" does not necessarily resolve the matter. We might assume that science speaks a universal language of "truth," but it does not. Khachatourians posits that science moves from unknown causes and effects to known causes and effects and back again, as the theory, methodologies, and evidence varies, with the result that there are few absolutes in science. Scientific knowledge is also contested, especially in such complex domains as human health. Citizens often ask questions to which science can have no answers, which simply highlights that scientific risk assessments are often forced to make implicit value judgements in order to come to a conclusion. Even when regulators use formal cost-benefit analysis, which involves explicit valuation of social impacts, the results must necessarily depend on a subjective valuation of things like human life and the environment. Moreover, it is not clear that there is a consensus about the science. Views often differ between countries, making it difficult for the international science community to agree on acceptable tolerances, the tests to be done, and how they should be interpreted. While the community of science may not respect national boundaries, it does respond to the questions that get asked, which then leads to competing scientific views. Fundamentally, what differs between countries is how we weight the information provided and how we balance competing interests, as, for instance, between consumers and producers or between human health and the environment. Some policymakers resort to using the precautionary principle as a benchmark when confronted with ambiguous scientific advice. Even if policymakers and regulators decide based on accepted science, many consumers do not trust their own government's, let alone a foreign government's, scientific judgement to adequately protect the safety of their food supply.

## TRADE

The characteristics of food are only a governance issue when a product enters a market — if food is not being bought and sold, there is no risk to manage. Initially, when buyer and seller knew each other, they could internalize risk analysis within their relations. Governments developed as an intermediary when markets became larger and as transactions took longer to complete in space and time. Globalization is the latest stage in straining such close relationships. We are observing massive change in the industrial organization of food. Millions of small farms and stores are still engaged, but seed and other input suppliers are concentrating, many processors operate on a global scale, wholesalers and distributors are growing, and some retail chains are now enormous. When the fragmentation of production, distribution, and consumption extends beyond the domain of any one state, governments are challenged both to understand the process and to ensure that the concerns of citizens and consumers are met at all stages of the process, without unfairly disadvantaging producers. This then is the challenge for risk management, defined by Codex as "the process, distinct from risk assessment, of weighing policy alternatives, in consultation with all interested parties, considering risk assessment and other factors relevant for the health protection of consumers and for the promotion of fair trade practices, and, if needed, selecting appropriate prevention and control options."

Food safety is a dramatic example of the regulatory difficulty states face in the era of globalization. As Phillips notes, technological change creates new products faster than our collective ability to assess their implications while new forms of transportation and expanding markets allow these products, and related pathogens, to move rapidly around the world because of the ever-increasing exchanges of goods and services in the global economy. What is interesting about "trade," therefore, is that it is the vector that brings pathogens and practices (e.g., regulations and laws) from different places into contact. Established products may not be understood in the new market, or may pose novel hazards for consumers based on different levels of exposure. New products may face regulatory review, but citizens are concerned about distant practices that may not be up to their standards. Information can be disseminated rapidly, but consensual knowledge does not keep up. Decisions with regulatory implications may be effectively taken within gigantic multinational firms, or within such diverse international organizations as the FAO, the WHO, the International Organization for Standardization (ISO) or the WTO, or at times may be preempted by civil society organizations. Regardless of where the decisions are made, the food safety system to be effective must take account of local farming practices, slaughterhouse practices, processors' quality control regimes, distribution systems, private standards, and labelling requirements, which makes it difficult for any one group to dictate to the system.

Phillips and Falkner examine the international institutions and show that international cooperation is limited by different national views about whether trade is more important than the environment or health. Depending on one's perspective, one would look to the WTO, the Convention on Biodiversity, the FAO or WHO. Other linked issues — including intellectual property rights in seeds and the approvals process for new pharmaceuticals — compound the complexity of international discussions. National and international provisions for labelling for food are often the ultimate battleground for these different perspectives. Some labelling regulations provide for eco-labelling (e.g., certifying that wood came from sustainably managed forests) while agreements on food inspection often set precedents for general principles under the WTO Technical Barriers to Trade (TBT) agreement (e.g., mutual recognition of testing for conformity to product standards). MacKenzie's contribution highlights the complex and arduous process required to establish standards through the *Codex Alimentarius* system.

Citizens demand direct regulation from governments while consumers implicitly drive regulation as producers respond to market signals. In these circumstances, voluntary labelling may be the best way to regulate certain food attributes. In theory this possibility is important because in such a diffuse industry, no coercive form of regulation could ever catch everything going on in the food system. Governments, however, worry about concentration in the production, distribution, or retail systems that could limit the competition that forces producers to be responsive to consumers.

While nation-states tend to be the only locus for making effective public policy, the food industry is now global. Craddock and Hepworth offer industry-based observations on the challenges facing global firms operating in this new environment. Countries have very different traditions and infrastructure for application of food regulatory regimes. Some countries use a market-based approach while others have an interventionist approach to food inspection and consumer protection generally. Now increased trade flows are exposing the problems with purely national approaches to regulation. Globalization brings national regulatory frameworks into conflict with each other, sometimes mediated by international regimes. New rules can disrupt trade, or work at cross-purposes to rules elsewhere. Phillips shows that there are currently six international institutions and various regional organizations working to manage risks in the international food system while at the same time keeping trade links open by managing disputes. The International Plant Protection Convention, Office international des épizooties, OECD, WTO, BioSafety Protocol, and *Codex Alimentarius* all have key regulatory roles. In addition, the international agri-food research and policy systems (FAO, WHO, the Consultative Group for International Agricultural Research, and the Rockefeller Foundation) are key agencies involved in building research and regulatory capacity in countries and regions. While they provide some support, it is not clear how these agencies could most usefully work with each other and with national governments.

## SAFETY

Everyone is aware that food-related issues have a much higher political profile than they did a generation ago. People in a great many countries are worried about transgenic foods, or GMOs. They worry about food additives that may be carcinogenic and about pesticide residues on fruit. New foods, such as new varieties of rice, might be essential for enhancing food security, but possibly at the cost of decreased biodiversity. In this book we have characterized the safety point of the triangle as being about risk communication, defined by Codex as "the interactive exchange of information and opinions throughout the risk analysis process concerning risk, risk-related factors and risk perceptions, among risk assessors, risk managers, consumers, industry, the academic community and other interested parties, including the explanation of risk assessment findings and the basis of risk management decisions."

Henson, Mann, and Powell each highlight that public perceptions have changed in recent years because of increased access to information, greater awareness of food-borne disease, and — especially in Europe — reaction to food disasters like BSE. We are all both citizens and consumers, which complicates risk management because, as mentioned earlier, what citizens say they want is not always consistent with the revealed preferences of consumers in the market. Public concerns go beyond human health, the central responsibility of the food safety system, to include such consumer issues as food quality (e.g., taste, appearance, and freshness) and such citizen concerns as ethics (e.g., animal welfare), biodiversity, and environmental sustainability. That is, where consumers are concerned with product characteristics, citizens are concerned with the thorny issues of how a product was produced. Our perceptions of food safety are based on our judgement about the relative value of these concerns. No food is absolutely safe for human health or for the environment or free of social concerns about its production methods. Hence, while food safety has a material basis, it is fundamentally a social construction, as we see most obviously with GMOs.

The local community tends to be the political space in which opinion about what is safe to eat is shaped, but information flows globally. We have seen a proliferation of non-governmental or civil society organizations (NGOs) that represent both consumers and citizens on issues including health, the environment, development and equity. Some are adjuncts to government or industry while others are critics of experts, producers, and officials. NGOs can be small, concerned with local issues, or large multinationals in their own right (e.g., Greenpeace and Friends of the Earth), concerned with global problems. These new forms of political engagement are being driven in Canada and the UK by declining trust in governments and in "experts." Differential patterns of NGO activity are explained in part by past disappointments that undercut public trust in food safety institutions.

Industry and government are challenged to respond to the changed demand for and supply of risk communications. Powell examines risk communications among producers in the supply chain and between producers and consumers through direct retail sales while Humphries discusses how the retail sector in the UK has responded to regulatory and consumer pressures for more communications. Mann also offers some observations about how industry and the new Food Safety Agency in the UK have begun to communicate to consumers about food safety.

## APPLYING THE TRIANGLE

One of the general phenomena associated with globalization is rising concern about the ability of social institutions to assess risks, manage risks, and communicate about risks. The authors of the chapters in this book examine each of the points of the triangles and show how science, industry, and governments have responded to the new pressures in the global food system.

As you read the chapters, keep in mind a number of key questions.

When do we expect governance to be necessary? When the science is clear, information is readily available and markets work, there would appear to be little need for formal governance structures. By induction, one could assume that when the science is uncertain, information incomplete and markets imperfect, governance may be needed.

Is there an optimal market or polity size for the functions at each point of the triangle? Risk assessment is based in the global community of science while risk management is rooted within the territorial state and risk communication may best be located within the local community. Despite the globalization of industry and information, the three poles of the triangle remain relevant, but none of the domains is self-contained. For example, risk, which is usually characterized as a measure of hazard times exposure, involves global scientific assessment but also national or local assessments on exposure. Hence, each of these areas spills over into others. There is significant ambiguity about when assessment stops and management and communication begins, which complicates determining the optimal location for each function.

How do the conventional tasks of risk analysis change when borders intervene? Some systems are producer-driven and others are consumer-driven. The same factors in different systems can result in different decisions.

When will governance be controversial? Different countries handle the balancing act of risk analysis differently, which will inevitably lead to demands for reconciliation or harmonization of the different approaches. Given the fundamental importance of food to people and to societies, this will inevitably lead to controversies. The challenge is to be able to predict when and where those controversies will occur.

When can international efforts yield substantive results? We have a long history in the international scientific community, in the international trading world, and in the social policy domain of seeking international solutions to problems facing multiple countries. International constraints on domestic policy can be helpful or disruptive. They can be explosive where the domestic standard is perceived to be higher or can be a non-issue where the international standard is higher. The authors examine a number of food issues that have been the subject of international discussion or negotiation.

There are no definitive answers to these questions, but the authors of the following chapters provide helpful insights into how risk analysis will change as food production and food governance respond to globalization.

NOTES

1.  See the *Codex Alimentarius* Commission "Definitions of Risk Analysis Terms Related to Food Safety." Risk analysis itself was defined as: "A process consisting of three components: risk assessment, risk management and risk communication."

# Science: Risk Assessment

# 2

# How Well Understood Is the "Science" of Food Safety?

*George G. Khachatourians*

How do scientific controversies affect national and global governance? What are the implications of changing scientific understanding for risk assessment? This book as a whole is premised on a triangular tension between science, safety, and trade, but we can see similar tensions within the domain of science itself. Without understanding these tensions, we cannot hope to understand the problems that new science creates for risk assessment, management, and communications.

In this chapter I discuss the policy issues posed by scientific change through the two topics that have generated the greatest public reaction: the use of antibiotics, hormones and wasted animal proteins products in meat production from animals, and the introduction of genetically modified organisms in food production from plants. Both forms of technology are aimed at increasing farm productivity. The use of antibiotics and hormones in meat-producing animals is intended first to enhance the animal's ability either to gain weight or to produce milk, and second as a prophylaxis against infectious disease, while the use of waste products was an attempt to improve the profitability of the sector. Plant biotechnology, an umbrella term for a number of interconnected disciplines, offers the promise of generating new products and processes that will make more efficient use of resources of land, water, and nutrients. The governments of both Canada and the United Kingdom embraced biotechnology as part of their national strategy in the late 1970s (Canada. Federal Task Force on Biotechnology 1981; Joint Working Party 1980).

The challenge facing scientists, regulators, industry, and citizens is that scientific understanding is continually evolving. This chapter uses Thomas Kuhn's approach to scientific development to illustrate how the science of food is changing, and how those changes have stressed the food safety system.

## SCIENTIFIC FRAMEWORKS

Risk assessment is defined as "a scientifically based process,"[1] but what might that mean? The risk-assessment process engages academic and government experts who collaborate in the generation of a consensual understanding of the issues with testing organizations, national expert panels, Food and Agriculture Organization (FAO)/World Health Organization (WHO) expert committees, international scientific bodies such as the International Plant Protection Convention (IPPC) and international organizations like the Organisation for Economic Co-operation and Development (OECD). While this process depends upon open dissemination of scientific information, it becomes more complex as disciplinary expertise fragments and specializes, interdisciplinary contributions to food issues increase, and as new products create new hazards and differing levels of exposure. Consumers and citizens respond to this fragmentation of information with growing distrust of experts and regulators. Scientists respond through a process of public debate, which can at times clarify and comfort or at times confuse and heighten concerns.

Thomas Kuhn (1970), in *The Structure of Scientific Revolutions*, attempts to construct a generalized picture of the process by which a science is born and undergoes change and development. He used the term "paradigm" to denote the body of knowledge that was part of "normal science." Scientific development, according to Kuhn, begins with the study of natural phenomena, which leads to a set of theories that explain differing viewpoints. This in turn results in the development of a pre-paradigm (or accepted body of knowledge), which after much empirical testing, evolves into a paradigm. A paradigm thus has a number of questions and answers that are accepted as "knowns." They are related to each other through the concepts and methods employed in research and are disseminated through publication in scientific journals. The evolutionary process of getting to normal science involves a series of research efforts and discoveries within the existing paradigm that lead to the further articulation of the paradigm, the exploration of other possibilities within the paradigm, the use of existing theory to predict new facts, the solving of scientific puzzles, and the development of new applications of theory. During the course of scientific inquiry, new phenomena are discovered and new or revised explanations are developed for these phenomena. In other words, new questions emerge when we discover something unexpected and new answers are found through further research. In both cases things that were "unknown" drive the search for answers, or new knowns. As research progresses, there may be further discoveries of natural phenomena that violate the paradigm-induced expectations that govern normal science. Researchers discover problems not previously known, and existing theory sometimes proves unable to account for the anomalous facts. Sometimes in such circumstances a researcher is able to define and exemplify a new conceptual and methodological framework incommensurate with the old one that leads to a new solution, thereby allowing the continuation of normal science within a new paradigm (Green 1971).

While Kuhn's examples of the formation and transformation of paradigms were drawn entirely from the history of the physical sciences, this analytical framework can be applied to food science. The challenge for risk management is distinguishing between situations of normal science, where a science-based approach to risk assessment can and should be left to experts, and the more ambiguous situation where unknowns make a "precautionary approach" the prudent course of action. The classic quadrant box can be used to describe the relationships between knowns (Ks) and unknowns (UKs). As detailed in Figure 1, within the frame of understanding the relationship between the known (K) and unknown (UK) questions and answers of normal science, four situations can arise. We can divide the frame into four combinations or boxes, where both the question and answer are known, (K–K), where one is known but the other is not, (K–UK, UK–K) and finally where neither the question nor answer is known (UK–UK).

FIGURE 1: Kuhnian Paradigms of Known and Unknown

|  | | Answer | |
|---|---|---|---|
|  | | Known | Unknown |
| **Question** | Known | K–K | K–UK |
|  | Unknown | UK–K | UK–UK |

The top left box provides few difficulties for scientists, regulators, and citizens as they consider food safety — problems for governance usually do not arise when the science is clear, information is readily available, and the community of science agrees that both questions and answers are "known." But scientists usually do not agree on what is known, so the boxes at lower left and upper right represent the common challenge for risk analysis, one that increasingly worries citizens who do not know which scientist, regulator, or activist to believe. The most dangerous situation is when both question and answer are not known, for then we are unaware of the risks we are running.

## APPLYING THE KUHN MODEL TO SAFETY ISSUES IN CATTLE PRODUCTION

Numerous food safety examples could fit this framework, but for this chapter I use four examples from the cattle industry (Figure 2) — the cases of antibiotic resistant bacteria (ABRB), the use of hormones in milk cattle, the presence of bovine spongiform encephalopathy (BSE) in cows, and the occurrence of variant Creutzfeldt-Jakob disease (vCJD) in humans. The first two cases rest on the foundations of microbiology and food science developed over the past 50 years. In the second two cases, however, normal science did not expect the emergence of prion diseases like BSE, which are still poorly understood, nor did it then expect that such diseases could move from one species to another, as is thought to be the case with vCJD.

FIGURE 2. Kuhnian Paradigms of Known and Unknown Related to Cattle Production

|  |  | **Answer** | |
|---|---|---|---|
|  |  | Known | Unknown |
| **Question** | Known | APRB | rBST |
|  | Unknown | BSE | vCJD |

In the first case, after 30 years of animal husbandry experience in the United Kingdom and Canada, we understand the questions about whether or not to augment animal feed with antibiotics and hormones for weight gain and meat quality. Today, most animal producers, and all intensified livestock operations (ILOs), use feed additives. From the standpoint of food production, both the problem (how to satisfy market demand while achieving profitability) and the solution (large-scale animal production) are known. ILOs utilizing hormone and antibiotic feeds experience a 5 to 14 percent improvement in weight gain. Over the past 30 years, the experience of using antibiotics has produced a number of new questions and answers from the standpoint of human and animal health, including the recent emergence of antibiotic resistant bacteria and their transmission throughout rural and urban environments. Vancomycin, an important antibiotic that has been used since the 1950s for humans and for farm animals since the 1980s, is similar to the antibiotic avoparcin, which is used solely in veterinary practice and animal feeds in the European Union (avoparcin has not been licensed for use in Canada or the United States as a feed additive because of the determination it had carcinogenic potential). Normal science agrees that, over time, bacteria can develop resistance to antibiotics, which requires the continuous search for new drugs; what was not clear was that resistance could transfer between animals and humans. In Europe, vancomycin-resistant enterococci (VRE) have been isolated, first from sewage treatment plants in Britain and small towns in Germany, and later in manure samples from pig and poultry farms (Khachatourians 1998). It is known that VRE isolates generally are cross-resistant to avoparcin (i.e., bacteria that are resistant to avoparcin will also be resistant to vancomycin). A possible reason for the presence of VRE in humans is that VRE is associated with animals being fed avoparcin. Accordingly, the response of normal science to the emergence of the new known factor, antibiotic-resistant bacteria, was a new known answer: a ban on avoparcin. Although the feed additive manufacturing industry in Europe initially protested the withdrawal of avoparcin from farm animal use, there are now restrictions on the use of avoparcin in Denmark and Britain (ibid.). While one might consider this a "known-unknown" situation, it is fairly clear that normal science anticipated this possibility — 30 years earlier the Swann Committee (1969) in the United Kingdom, fearing that there were potential problems had recommended strict adherence to regulations governing the use of antibiotics in animal feed — and none of the normal science needed to change to adapt to the new circumstances. Scientists and regulators simply extended regulations to manage this new circumstance: the United States, Sweden, and Denmark followed suit in 1997, the WHO in 1998, and several Canadian provinces more recently (ibid.).

The second case involves the use of recombinant bovine somatotropin (rbST), a non-therapeutic drug produced by genetic engineering, to increase milk production in dairy cattle. The question is known (i.e., how can milk production be increased safely), but the answer is not (i.e., scientists continue to debate about the long-term effects of rbST on animal and human health). In 1994, both the European Union and Canada imposed a moratorium on the use of rbST, although BST occurs naturally in cows. From 1993 to 1995 inquiries were held to resolve the scientific and policy aspects of this issue. On 5 May 1998, the Senate of Canada unanimously passed a motion urging the government to defer licensing rbST for at least one year and thereafter until the long-term risks to public health were known (Canada. Standing Senate Committee on Agriculture and Forestry 1999). On 14 January 1999, Health Canada announced that it would not approve rbST for sale in Canada. The disapproval was based on the findings of two expert advisory panels, both of which recommended that approval for rbST not be granted until long-term studies on human health were submitted and reviewed.

The third case concerns the emergence of BSE in the cattle population in the United Kingdom in the 1980s. In this case the question (what causes BSE?) was known too late, but the preliminary answer was readily apparent (cattle became visibly affected in the mid-1980s). Normal science had part of the question — the presence of scrapie in sheep had been known for decades and there was scientific knowledge of kuru-kuru and anthropological records of neuro-degenerative disease due to certain tribal cannibalistic practices (Rhodes 1997) — but the link to the problems facing the cattle industry in the 1980s was not clear. The outbreak of BSE did not on the face of it appear to be solely linked to the diet of animal proteins. English herds appeared to be especially at risk while Scottish cattle and North American cattle were not. The question began to be formulated with the discovery of prion diseases and the correlation between infected animals and the eating of proteins by animals of the same or similar species. The different incidence of the wasting disease in cattle between England and other areas that also allowed a diet of animal proteins led to a further discovery that prions could be reduced or eliminated if the materials were heated to a high temperature in preparing the meal. The United Kingdom had lowered the required temperature years earlier while the other jurisdictions maintained a higher required temperature. Even before the different bits of evidence and theory helped the scientists formulate a complete known question, governments in the United Kingdom and North America acted in a precautionary way and changed ruminant feeding regulations to ban animal proteins or require more stringent processing. The United Kingdom, as early as 1988, tightened rules on feeding animal proteins. Even so, in the absence of a clear question, regulators in continental Europe continued to allow feeds with animal proteins while others allowed the import of potentially infected cattle from the United Kingdom. As a result, BSE gained a foothold in cattle herds in France, Germany, and the Low Countries in the late 1990s, causing a tightening of rules there as well. Thus, no major cattle producers now allow industry to feed cows with protein materials derived from other ruminants. The initial incidence in the United Kingdom of BSE, its epidemiological rise in the cattle herds and the scientific cause-effect track were finally known in 1998, restoring the management of animal feeding to the known-known world, as fully fleshed out questions and answers are well on the way to becoming part of normal science.

In the fourth case, both the questions and answers were unknown, since nobody thought to ask if mad cow disease could spread to humans. This UK–UK situation initially created a false sense of security and, then, when evidence of some risk appeared, led to widespread concern in the United Kingdom. At the same time as the cattle industry was sorting out what BSE was and how it was transmitted in the cattle industry, epidemiologists noted a troubling rise in the number of people affected with Creutzfeldt-Jakob disease (CJD), a rare and fatal neuro-degenerative disease of unknown cause. This, therefore, represented a case of an unknown question (what is the link between BSE and CJD) and an unknown answer (is the incidence of CJD a problem related to BSE). Because of the epidemic of BSE in cattle, in May 1990 the United Kingdom established a national surveillance program for CJD in the country. By 1996 the unit was able to report (Will *et al.* 1996) that some new cases of CJD had neuropathological changes which, to their knowledge, have not been previously reported and that those affected were relatively young people (the usual case of sporadic CJD in the United Kingdom affected individuals who were aged 50 to 70). While the program determined that the average number of cases of sporadic CJD identified annually after 1990 was higher than in previous surveillance periods extending back to 1970, it was impossible to say with certainty to what extent these changes reflect an improvement in case ascertainment and to what extent, if any, they reflect changes in incidence. Up until 31 December 2000, there were 84 deaths from definite or probable variant CJD (vCJD) in the United Kingdom (in addition two probable cases died in January 2001 and a further seven probable cases remained alive as of 31 January 2001). Of the 84 deaths to 31 December 2000, 75 were confirmed neuropathologically with a further two awaiting confirmation. Over time this unknown-unknown situation has shifted slowly to a situation closer to normal science. The surveillance program produced evidence that something unusual or unexpected was occurring, and spurred on scientists to seek out a better description of the problem (answer) and then to seek causes of the vCJD (the question). The difference in this case is that in the absence of at least one known, the government had no way to react. Hence, although scientists had suspicions that vCJD might be linked to BSE, their inability to establish that link until 1996 forestalled any government response. In 1996, an article in the *Lancet* (Will *et al.* 1996) formally defined the answer (there was a new variant CJD) and pointed to a suspected causal link (with BSE and cattle fed with ruminant proteins), which provided some direction for government action. While the government response was not exemplary (see chapters by Patricia Mann and Catherine Humphries), it signaled the reorientation of this problem from being unknown to at least being partly known.

The four-quadrant model of knowns and unknowns applied against safety concerns related to animal rearing illustrates a number of points. First, if cause and effect are known and part of normal science, assessing, managing, and communicating about risks is relatively straightforward. Where one of either the cause or effect is unknown, governments are placed in a tricky position of trying to find the best response, but they are often forced to react in some precautionary way. Where both cause and effect are unknown, blissful ignorance rules and governments are neither pressed to nor able to act. Science needs to continuously add some knowledge of either cause or effect for governments to be able to respond.

## BIOTECHNOLOGY AND FOOD

A different but equally significant issue in science policy discourse is the use of biotechnology in foods. In terms of science, genetic engineering of plants is not synonymous with biotechnology, as some choose to view it, but one ingredient of biotechnology. As a part of science, genetic engineering promises to make important contributions to food design and production, creating an agricultural framework that provides both environmental sustainability and food security (Khachatourians 2001). But in getting there, there are a number of discordant science and policy elements that must be addressed: the development and application of biotechnology for foods has precipitated a major debate within and beyond the scientific community about what safety means and how it can be measured and secured. As a new part of science, biotechnology is not part of normal science: scientists, regulators, and citizens disagree on the questions and on whether science is approaching the answers. That is, few people think that we are in a K–K situation, and many worry that we are really in the UK–UK situation of not knowing what we do not know about the new science and its applications. The effort to define the unknowns and to search for new knowns has triggered a wide debate about how science can and should support governance systems.

Biotechnology definitely has ample "unknown" areas and the paradigms about how to gain insights are shifting rapidly and often inconsistently. In both the United Kingdom and Canada, for instance, the public has been highly selective in accepting commodities derived from engineered micro-organisms — health-care products derived from transgenic plants go uncontested while genetically modified (GM) foods are often spurned. As a result, the debate over GM foods has divided both the scientific and policy communities as much as the general public. Safety is no longer simply a scientific concept, it now has a major societal component. Policymakers face an interpretative dilemma. Governments on both sides of the Atlantic, acting more from cultural and societal perspectives than from a scientific base, have used the break in normal science to adopt inconsistent and often conflicting precautionary positions related to the use of biotechnology and transgenic techniques.

There are many questions that science alone cannot adequately address. Where do the biological facts meet social truths? Are social responses to advances in food sciences influenced by normal science, experimentation and observation, or paradigm shifts? As pointed out by Sackett *et al.* (1996), can and should compilations or critical reviews of the science literature provide authoritative summations?

How does a scientist examine re-interpretation and conjectures? These questions go well beyond the scope of normal science but are fundamental to the effective governance of food.

Finally, while objectivity in performing scientific inquiry is a must, the biotechnology debate has starkly revealed that scientific experts do not transmit objective information to policymakers in a way that will have a positive influence on the formulation of policy. Scientists by and large do not articulate scientific data in a manner that is comprehensible for politicians, leading to the inevitable absence of objective data in policy debate. As a result, emotion and rhetoric are often more influential than objective data. Furthermore, research programs that attempt to address public concerns often have the opposite effect to that promised. During times of breaks in normal science, politicians and scientists participate in consensual and mutually aggrandizing promises and predictions, often offering a cure for cancer here, and the elimination of environmental pollution there. Calls for further research reflect a bias about the perceived role of science in policy-making. The prevailing view is that science is there to solve problems. Instead, one could argue using Kuhn's framework that science is perhaps more important in defining the unknowns and seeking new knowns, thereby creating the foundation for sound governance. Without knowledge about the problems and answers, risk assessment, management, and communications will founder.

## LESSONS FOR GOVERNING FOOD

Advances in food sciences including biotechnology and particularly the techniques of genetic engineering have generated new hopes and fears about food safety. These interests have attracted some of the best and brightest scientists to these new sciences and their application toward producing safe foods. The potential of new knowledge being applied to understanding issues of food safety is enormous. It is critical, however, for both scientists and the rest of society to understand how normal science operates and how science can contribute to defining the knowns and unknowns that can influence effective governance of food safety.

Nevertheless, one must be realistic about what science can do. The appeal to "science" will not necessarily resolve disputes. We might assume that science speaks a universal language of truth, but it does not. Scientific knowledge is especially contested in such complex domains as human health. Citizens often ask questions to which science can have no answers, which simply highlights that scientific risk assessments often are forced to make implicit value judgements to come to a conclusion. Even when regulators use formal cost-benefit analysis, which involves explicit valuation of social impacts, the results must necessarily depend on a subjective valuation of things like human life and the environment.

Moreover, it is not clear that there is such a thing as normal science in this world of rapidly advancing knowledge. Even if there were a consensus about the science, views about what matters often differ between the various societies and cultures, with the result that countries adopt divergent policies. As a result, the international science community is often unable to agree on acceptable tolerances, the tests to be done and how they should be interpreted. While the community of science may not respect national boundaries, it does respond to the questions that get asked, which leads to competing scientific views. Fundamentally, what differs between countries is how we weight the information provided and how we balance competing interests, as, for instance, between consumers and producers or between human health, the environment, and the economy. Finally, even if policymakers and regulators do decide based on some internationally accepted scientific consensus, many consumers do not trust their own government's, let alone a foreign government's, scientific judgement to adequately protect the safety of their food supply.

## NOTE

1.  See the *Codex Alimentarius* Commission "Definitions of Risk Analysis Terms Related to Food Safety." Risk analysis itself was defined as: "A process consisting of three components: risk assessment, risk management and risk communication."

## REFERENCES

Canada. Federal Task Force on Biotechnology. 1981. *Biotechnology: A Development Plan for Canada*. Report of the Task Force on Biotechnology to the Minister of State for Science and Technology. Ottawa. Supply and Services Canada.

Canada. Standing Senate Committee on Agriculture and Forestry. 1999. *rBST and the Drug Approval Process*. Report of the Committee, March. Ottawa: Supply and Services Canada.

Green, J.C . 1971. "The Kuhnian Paradigm and the Darwinian Revolution in Natural History," in *Perspectives in the History of Science and Technology*, ed. D.H.D. Roller. Norman, OH: University of Oklahoma Press, pp. 3-25.

Joint Working Party. 1980. *Biotechnology*. The report of a Joint Working Party (Advisory Council for Applied Research and Development, Advisory Board for the Research Councils, and The Royal Society) of United Kingdom. London: Her Majesty's Stationery Office.

Khachatourians, G.G. 1998. "Agricultural Use of Antibiotics, and the Evolution and Transfer of Antibiotic Resistant Bacteria," *Canadian Medical Association Journal* 159:1129-36.

_____ 2001. "Agriculture and Food Crops: Development, Science and Society," in *Transgenic Plants and Crops*, ed. G.G. Khachatourians, A. McHughen, W-K. Nip, R. Scorza and Y-H. Hui. New York: Marcel Dekker Inc.

Kuhn, T. 1970. *The Structure of Scientific Revolutions*, 2d ed. Chicago: University of Chicago Press.

Rhodes, R. 1997. *Deadly Feasts: Tracking the Secrets of a Terrifying New Plague*. New York: Simon & Schuster.

Sackett, D.L, W.M.C. Rosenberg, J.A. Muir Gray, R.B. Haynes and W.S. Richardson. 1996. "Evidence Based Medicine: What It Is and What It Isn't," *British Medical Journal* 312:71-72.

Swann Committee. 1969. *Report of the Joint Committee on the Use of Antibiotics in Animal Husbandry and Veterinary Medicine*. London: Her Majesty's Stationery Office.

Will, R.G., J.W. Ironside, M. Zeidler, S.N. Cousens, K. Estibeiro, A. Alperovitch, S. Poser, M. Pocchiari, A. Hofman, P.G. Smith. 1996. "A New Variant of Creutzfeldt-Jakob Disease in the UK," *Lancet* 347:921-25.

# Trade: Risk Management

# 3

# Food Safety, Trade Policy and International Institutions

*Peter W.B. Phillips*

## INTRODUCTION

The provision of adequate supplies of relatively safe, affordable, and nutritious foods to feed a rapidly growing world population and increasingly more affluent consumers has been one of the key accomplishments of the twentieth century. As recently as the nineteenth century most people faced a life of malnourishment due to limited food supply with all too frequent risks of illness caused by unsafe food. While pockets of malnourishment and risk continue in places, much of the world now has adequate quantities of safe food. This transformation included the development of a food supply system that involved scientific advancement, evolving regulatory capacity, and industrial and market development. That effort began with largely domestic capacity-building, but rising demand for food in this century caused international trade to become an increasingly important precondition for food security. An extensive array of international agreements and institutions were developed during the twentieth century to facilitate trade while at the same time ensuring food safety.

The increasing sophistication of detection methods, the emergence of complex diseases (e.g., bovine spongiform encephalopathy [BSE]) and the advent of new commercial technologies (hormones and genetically modified foods) has effectively destabilized the postwar consensus that underpins the international regulatory system, precipitating the re-nationalization or privatization of parts of the regulatory system. The international trade policy system is currently grappling with finding new international consensus on how to manage new risks and new concerns related to these developments. This chapter offers an assessment of the evolving issues and structures, with a focus on food safety issues of particular interest to Canada (as a significant food exporter) and the United Kingdom (as a major food importer).

The rest of this chapter provides an overview of the relationship between food safety and international trade rules. The next section offers a brief comment on the historical relationship between trade and food safety. Section three examines the relationship between food safety risks and global production, trade, and consumption. The fourth section reviews the international rules and institutions tasked with securing food safety and facilitating international trade. The following section examines how the private sector is responding to the new international market uncertainties. The final section offers some concluding comments and observations on strategies for resolving the current difficulties in the regulation of safe food trade.

## THE SCIENCE, REGULATION AND TRADE OF SAFE FOODS

Even a cursory review of the history of food safety and international trade shows that science and technology, regulatory capacity, and market structures are the three pillars required to support the delivery of adequate quantities of safe food. Furthermore, past experience has shown that when one of three pillars for safety lags, trade disputes arise and risks proliferate.

Food safety has been a concern of governments from the earliest times: Egyptian scrolls identify the labelling applied to certain foods; in ancient Athens, beer and wines were inspected for purity and soundness; the Romans had a state food control system to protect consumers from bad produce; and during the Middle Ages, individual European countries passed laws concerning the quality and safety of eggs, sausages, cheese, beer, wine and bread, some of which still exist today (Codex 2000). In each case the systems functioned because new technologies (e.g., brewing, fermenting, and processing) were matched with commercial institutions and effective regulatory capacity.

The second half of the nineteenth century saw the beginnings of the modern food safety system. As food chemistry developed it provided the capacity to evaluate the "purity" of a food and to detect the use of hazardous chemicals in food. New transportation systems enabled a significant increase in international trade in foods: cereal shipments between North America and Europe grew markedly in the late 1800s while the first international shipments of frozen meat occurred between Australia/New Zealand and the United Kingdom. The first general food laws were adopted in key markets and basic food safety systems put in place to monitor compliance. In the Austro-Hungarian Empire between 1897 and 1911, for example, a collection of standards and product descriptions for a wide variety of foods was developed as the *Codex Alimentarius Austriacus*. Although lacking legal force, it was used as a reference by the courts to determine standards of identity for specific foods (Codex 2000).

These early food safety systems were national in scope but some were exported through the market access rules of the large importing regions. Indeed, as international food trade increased, international trade agreements became an integral part of the modern food safety system. This began in the early 1900s when a number of food trade associations sought to increase world trade by developing harmonized standards for specific products. In 1903, for instance, the International Dairy Federation developed new international standards for milk and milk products and others followed suit. Over the intervening 50 years while commodity groups pushed for international food safety rules to facilitate trade, governments focused most of their attention on establishing rules and systems to protect their domestic industries and producers.

The Great Depression and then World War II caused a shift toward consumer concerns. First, the development of more sensitive analytical tools and knowledge about the nature of food, its quality and associated health hazards increased consumer interest in food safety. Second, the leading trading nations moved to the fore, aggressively seeking more liberal trade rules and international institutions. In the area of food safety, this involved the creation of the Food and Agriculture Organization (FAO), *Codex Alimentarius* and a series of international trade agreements under the General Agreement on Tariffs and Trade (GATT) and more recently the World Trade Organization (WTO). (These institutions are discussed in detail later in the chapter.)

Even as international rules were finally coming to the fore in the past decade, the three pillars of food safety have become detached. Science advanced rapidly, both allowing for the detection of new risk factors (prion diseases) and introducing new technologies with new concerns (hormones, genetically modified organisms). Partly due to incomplete detection systems and partly due to inflexible structures, the marketplace has been unable to segregate the risky from less-risky produce. Regulatory systems, both domestically and internationally, have tried but have been unable to keep pace with developments in either science or the markets.

## FOOD SAFETY AND INTERNATIONAL TRADE

Food safety is fundamentally an international challenge, as many of the basic food commodities are extensively traded and a significant share of the whole and processed foods we eat incorporate some imported foodstuffs. This involves a large number of developed and developing countries both on the production and consumption end of the supply chain. The actual incidence of risk therefore is a function of both the characteristics of the food and the dispersion of food through trade.

Risks related to specific foods are inherently a function of the type and nature of the foods and the systems used to produce, process, market, and label the goods we consume. Plant products, for instance, could be contaminated by impure seeds (rapeseed seeds would contaminate canola grade seed), by chemicals that leave a residue in the product (herbicides and insecticides) and by abiotic or viral stresses that cause microbial or nutritional changes (aflotoxins in maize). Poor farming practices could also affect product quality — for example, improperly hilled potatoes could cause part of the tuber to turn green, indicating the presence of solanine (McHughen 2000). Most recently, the introduction of new technologies — genetically modified organisms (GMOs) — has raised concerns about new allergenic, toxic or subtler long-term cumulative effects. Moving downstream, manufacturing processes could introduce or magnify the presence of toxic microorganisms or other adulterations. Finally, transportation, wholesale, and retail systems could lead to wastage, blending (nuts into other ingredients), contamination or mislabelling (presence of nuts) that would create new risks. Similarly, the quality and safety of animal products could be affected by impure breeding (chickens), contaminated feed inputs (scrapie-infected meals), antibiotic or hormone residues (bovine growth hormone in beef) and poor production processes (needles remaining in carcasses). Recently there has been increased concern about how genetically modified (GM) feeds might affect food safety. Even more significantly, the processing stage has great potential to introduce micro-organisms that could cause contamination and food-borne illnesses. Finally, given the higher rates of spoilage for processed meats and animal products, small failures in the food distribution system could lead to major food safety concerns.

The dispersion of primary and processed products also influences potential risks. Almost every major staple food and many processed foods are now extensively traded globally, with the result that consumers are at risk from both local production and distribution risks but also from imported risks. Table 1 shows that for almost every major foodstuff, international trade accounts for both a significant and rising share of domestic supply. The results range from sugar, which has effectively no significant international trade due to national farm policies, to fish, with 35 percent of the global production crossing international boundaries. Given that international trade rules bind governments to real commitments on market access for processed food, trade is an even more significant share of many processed food markets. Canada, for instance, exports and imports annually at least as much processed food as commodity foodstuffs.

Modern biotechnology, which many date from the early 1970s, is the latest and, perhaps, most fundamental innovative technology to be applied to the agrifood sector. The array of new tools and techniques called biotechnology — including genomics, tissue culture, micro-propagation, cloning, marker-assisted

TABLE 1: Global Production and International Trade in Key Foodstuffs, 1961–1998

| Products Ranked by Volume of Global Production | Global Production (millions of tonnes) | Trade Dependence (Exports/production) % | |
|---|---|---|---|
| | 1996–98 | 1961–63 | 1996–98 |
| *Crops* | | | |
| Cereals | 1,891 | 10 | 14 |
| Sugar crops | 1,507 | <1 | <1 |
| Vegetables | 608 | 3 | 6 |
| Oil crops | 436 | 13 | 13 |
| Fruits | 432 | 9 | 19 |
| Pulses | 55 | 3 | 14 |
| Spices | 5 | 12 | 22 |
| *Animal products* | | | |
| Milk and milk products | 549 | 5 | 12 |
| Meat | 251 | 6 | 9 |
| Fish | 120 | 31 | 35 |
| Poultry | 58 | 3 | 12 |
| Eggs | 51 | 4 | 2 |
| Animal fats | 30 | 19 | 23 |
| Cheese | 15 | 10 | 19 |

Source: FAOSTAT Agriculture Statistics, Commodity Balances.

breeding, gene splicing, and transgenes — now allow breeders to selectively modify plants or animals at the molecular level. Although the first unconfined release of a GM crop was tobacco in China in the late 1980s, the main effort to commercialize GM crops has been in the past five years. During 1995 a small number of acres of GM varieties of canola, cotton, maize, soybeans, and tomatoes were produced in Canada and the US. Since then seven more modified crops have entered the market: flax, melons, papaya, potato, rice, squash, and sugar beet. James (2000) estimates that global production of these crops, scattered about 13 countries, grew to approximately 109 million acres in 2000.

A number of countries have also approved the release of one or more varieties of genetically modified fish (salmon), trees (poplar), microbes (ice minus), homones (recombinant bovine somatotropin [rbST]) and various vaccines for animals. More than 40 other crops and a range of animal species and microbes have been genetically modified and await regulatory approval in various countries involved in the international food trade (OECD 2000*a*).

Specifically, with respect to GM crops, commercial production is concentrated. James (2000) estimates that the US planted 68 percent of the global area for transgenic crops in 2000, Argentina planted 23 percent, Canada 7 percent, China 1 percent and Australia, Bulgaria, France, Germany, Mexico, Romania, Spain, South Africa, and Uruguay together account for less than 1 percent of total acreage. In past years, Portugal, Japan, the Ukraine, and the UK planted some GM crops but were not producing in 2000. Soybeans accounted for 58 percent of the global acreage in 2000, maize 23 percent, cotton 12 percent, canola 7 percent, and the other crops less than 1 percent in total. In 2001, GM flax and potatoes were withdrawn from the market.

Output from that limited number of countries is consumed widely around the world through international trade (Table 2). Given that none of the countries producing GM varieties has completely effective crop segregation systems in place for those crops commercially produced, even small amounts of GM production could co-mingle with the general commodity stream and lead to diffusion of the seed through international exports. Although a maximum of eight countries produce and export any one of the genetically modified commodities, this production is concentrated in countries that dominate both world production and world trade in those products. For example, although GM canola is produced commercially in only two countries (Canada and the US), they together produce approximately 23 percent of global production and account for 50 percent of global exports. GM maize, soybeans and flax are produced in countries that account for more than 80 percent of world trade in those products. In contrast, a number of the GM crops are either not extensively traded or are produced in countries that are not major exporters (e.g., beets, papaya, potatoes, and tobacco).

Given that most of these GM products are commodities subject to significant blending and shipping, it is often hard to determine the exact originating country of all parts of any shipment. As a result, any GM production that enters the international market could potentially be imported by any of the importing countries. As noted in Table 2, up to 177 countries import some quantity of the 13 crops that have been modified. As a result, domestic and international trade rules, and ultimately private companies, are being forced to adapt to consumer and citizen concerns.

For sound commercial reasons innovators have commercialized these new GM foods in those countries with large production bases, which also tend to be the major exporting countries. Over time this uneven adoption of largely yield-enhancing technologies will increase the exportable surpluses in those countries, increasing the need for international market access. As a result, biotechnology

TABLE 2: Distribution of Production and Trade in Genetically Modified Food Crops, 1998

| Commodity | Total World Production | | Production in Countries Producing Approved GM varieties | | Exports | Imports |
|---|---|---|---|---|---|---|
| | No. of Countries | Volume (Mt) | Countries | Total % | Total World Exports Originating in Countries with GM Production % | Total No. of Countries Importing Commodity |
| Canola | 51 | 35.9 | Canada, US | 23 | 50 | 68 |
| Flax | 46 | 2.7 | Canada, US* | 46 | 81 | 74 |
| Maize | 158 | 613.3 | Argentina, Canada, France, Portugal Spain, South Africa, US | 53 | 85 | 168 |
| Melon | 17 | 0.7 | US | <1 | <1 | 13 |
| Papaya | 47 | 5.1 | US | <1 | 5 | 56 |
| Potato | 148 | 298.5 | Canada, Ukraine, US | 14 | 12 | 177 |
| Rice | 114 | 575.9 | US | 1 | 11 | 82 |
| Soybeans | 83 | 157.7 | Argentina, Canada, Romania, US | 81 | 88 | 114 |
| Squash | 83 | 14.8 | US | <1 | <1 | 32 |
| Sugar Beet | 52 | 259.6 | US | 11 | 19 | 26 |
| Tobacco | 128 | 6.9 | China | 36 | 9 | 154 |
| Tomatoes | 159 | 90.9 | US** | 37 | 54 | 140 |

Note: *Only seed multiplication was undertaken for GM flax in 1998; the only GM flax variety was withdrawn from the market in 2001.
**There are inadequate records to determine whether GM tomatoes are being grown in four other countries that have approved these varieties. Cotton is not included as it is not a food crop.

Source: Phillips (2001).

will add pressure for renewed rules for international market access in the agri-food sector.

## DOMESTIC REGULATION OF FOOD SAFETY

While domestic regulatory systems currently provide the foundation for regulating food safety, delivering safe food involves a wide array of government agencies, producers, processors, wholesalers, retailers and, ultimately, consumers. The OECD (2000c) has just completed a major review of national food safety systems and activities, which highlights the similarities and differences of the various domestic systems. At root, all of the countries in the report are somewhat similar in that they base their food safety decisions on a risk analysis approach, which is understandable in that these systems have evolved over the past 50 years in tandem with the development of international agreements on standards and procedures. The differences lie in how they implement their risk assessment, risk management, and risk communication.

Each of the member states in the OECD generally has some relatively well-defined set of criteria for evaluating risk and has a system for undertaking that review. All countries indicate that they use scientific risk analysis as part of the assessment. The US, at one extreme, emphasizes that their decisions are always based on objective scientific evidence and as such, they are subject to appeal through the courts. Other countries indicate that an array of more subjective, socio-economic criteria could be considered in the review of risks. In many cases it would appear that these other criteria are inserted at a more political stage of review (e.g., through the Article 21 Committee in the European Union). The evidence available suggests that the scientific judgements of risk by regulators in different countries tend to be very similar, whereas the political judgements have significant potential to vary. As for the process for review, most countries have an array of agencies and legal authorities involved. In the past there was a tendency to divide the responsibilities among different authorities, with separate acts and regulators for each stage of the food chain or for each type of food product. Increasingly, the trend is to develop more centralized authorities. The US has the Food and Drug Administration, Australia and New Zealand created the ANZ Food Agency in 1991, Canada created the Canadian Food Inspection Agency in 1997, the UK created the Food Safety Agency in April 2000 and Japan announced it will create a single agency by April 2001. Meanwhile, most of those countries have reviewed or are in the process of reviewing and consolidating their legislation. More problematically, all of these regulatory systems depend critically on the involvement of participants from the food system. As risks are often specific to the technologies and processes used, individual companies and sectoral associations are key suppliers of information that regulators can use to evaluate the safety of the food. Given constrained budgets and growing workloads, one challenge all governments face in assessing risk is that rapidly advancing science may be beyond the capacity of regulators to assess. National regulators may not have adequate resources to fully assess the particular risks of products submitted for review, which could either result in risky products entering the market or stall safe and beneficial products in the regulatory system.

Risk management tends to be less centralized in most countries. Given the federalist nature of many of the states, especially Australia, Canada, the EU, and the US, and given the practicalities of monitoring a wide range of foods across large areas, most countries have adopted extensive networks of assessors, auditors, and inspectors to maintain the safety of the food system. Most of the major exporting countries — Australia, Canada, New Zealand, and the US — indicated that they are also aggressively moving toward more industry involvement in food safety, through industry-based standards systems such as Good Manufacturing Practices (GMP) or Hazard Analysis Critical Control Points (HACCP) systems.

Risk communication is less organized (or at least less well communicated!). Most OECD countries identified that they have systems for communicating with consumers about food safety risks, but most of the systems do not appear to be formal or comprehensive and all depend on the commercial media to get their message out. While the increased use of the Internet for posting information and decisions is helping to increase the transparency of the system, more could be done. The US goes the furthest in committing to communicate during the risk assessment process, with its requirement to post draft decisions in the Federal Register. Other assessment systems are more opaque until decisions are reached, and even then the decision documents are often not detailed enough to assess critically the basis for the decision (Doern 2000 discusses this problem in Canada). Communications related to risks in the marketplace are equally challenging. Labelling is the primary formal vehicle for conveying information. While all countries require labelling for contents and for specific scientifically verifiable product risks, such as nutritional change in a product or presence of allergens, many of the countries do not even require labels for nutritional elements (e.g., Canada). Manufacturers voluntarily label a wide range of other food features, such as nutrition, country of origin, production method (e.g., organic) and other socioeconomic or ethical features (e.g., animal welfare, Kosher, Halal). Recently, Australia, the EU, Japan, and South Korea have announced or adopted mandatory labelling for GM foods (see Phillips and McNeill 2001 for details). Many regulators and food industry companies are trying a variety of other means of communicating with consumers, including point of sale information, but consumers still express confusion and uncertainty about what risks relate to the food they eat (Einsedel 2000).

The advancement of new technologies, specifically biotechnology, has forced many of the domestic regulatory systems to re-evaluate their risk analysis processes. As is obvious from the foregoing discussion, all OECD countries have some form of domestic food safety regulation that predates the arrival of biotechnology in the global production and trade system. Depending on the degree of concern in the domestic market, countries have either relied upon those systems, in conjunction with the regulatory oversight of the exporting country, or have built new or modified systems to handle their specific concerns about biotechnology. Two divergent regulatory approaches have evolved. The North American regulatory approach (generally used by Australia, Canada, Japan, Mexico, New Zealand, and the US) is relatively pro-supply, relying on a legalistic interpretation of mostly scientific assessments, while the European model pursues a precautionary approach that reflects citizen and political concerns (Isaac and Phillips 1999). Although both approaches delivered common decisions in early years, they have diverged significantly since 1998, with the result that more than 25 GM food products have been approved in the US and Canada that are not approved in the EU.

As a result of the incomplete and often conflicting domestic regulatory systems in both exporting and importing countries, the regulation of GM foods has become an international issue. There has been extensive effort on a number of fronts in recent years to try to resolve the impasse over the regulation and trade of genetically modified products.

## INTERNATIONAL REGULATION OF FOOD SAFETY

In the past, new food products that were reviewed and approved in the country of production were generally granted unrestricted access to global markets. Rising consumer concerns about food safety (Angus Reid 1999) and widely varying citizen concern about the environmental, social, economic, and ethical aspects of GM foods has led many countries to change their practices and require a domestic review of the food products before they are allowed to be imported or sold to local consumers. Governments in most exporting and importing countries acknowledge that these national reviews could adversely affect the free flow of trade in food products. As a result, there has been renewed international effort to find a means of redesigning the multilateral trade system to more comprehensively assess and manage food safety risks. In effect, each of the institutions is attempting to create greater consensus about how to manage food safety risks.

TABLE 3: The Current Array of Institutions Regulating International Trade in GM Crops

| Institution | Date | Coverage | Member States | DSM | Orientation |
|---|---|---|---|---|---|
| Office international des épizooties (OIE) | 1924 | Infectious animal diseases | 155 | Non-binding; sets WTO standards via SPS S.3.4 | Harmonize import and export regulations for animals and animal products through International Animal Health Code |
| International Plant Protection Convention (IPPC) | 1952 | Pests and pathogens of plants and plant products | 107 | Non-binding; sets WTO standards via SPS S.3.4 | International Standard for Phytosanitary Measures (ISPMs) involving quarantines |
| The *Codex Alimentarius* Commission (Codex) | 1962 | Food labelling and safety standards | 165 | Non-binding; sets WTO standards via SPS S.3.4 | International standards to provide guidance to the food industry and protection to consumer health |
| OECD | 1961 | Harmonization of international regulatory requirements, standards and policies | 29 | None | Consensus documents |
| Regional initiatives | 1990s | Harmonization of the science of regulation | Various groupings | None | Regional side agreements, MOU, MRA, formal dialogues, and joint research projects |
| WTO | 1995 | Trade in all goods and most services | 138 | Binding | Establish rules for transparency and dispute settlement through TBT and SPS agreements |
| BioSafety Protocol | 2001? | Transboundary movements of living modified organisms | 63 signed; 0 ratified (50 ratif. required) | None | Will require advanced, informed agreement for first shipments of LMOs intended for deliberate release; commodity shipments to be notified |

There currently are seven international bodies vying to coordinate and regulate different aspects of food safety (Table 3). Conceptually they represent a progression from institutions that are largely science-based (IPPC, Office international des épizooties [OIE], and Codex), one trade-based (WTO), and others that have broader objectives such as environmental protection and other social and political goals (OECD, regional initiatives, and the BioSafety Protocol [BSP]). This section offers a thumbnail sketch of each of the institutions involved in regulating trade in food safety and a brief synopsis of where the system is going and how they work together and with the national regulators (see Buckingham *et al.* 1999; or OECD 2000*d* for more details).

The IPPC and OIE are multilateral treaties that seek to protect plants and animals from the spread of pathogens through international trade. These treaties provide much of the scientific consensus that underlies domestic food safety systems. The IPPC protects natural flora, cultivated plants, and plant products and the OIE protects animals and fish. In collaboration with both regional and national plant and animal protection organizations, they provide a forum for international cooperation, harmonization, and technical exchange of plant and animal protection information. In terms of new technologies, the IPPC has addressed the international regulation of GM crops through several International Standards for Phytosanitary Measures (ISPMs) while the OIE has developed standards for diagnostic reagents, sera, and vaccines for animals in the International Animal Health Code. Both institutions have their own non-binding dispute avoidance and settlement systems, but their most important role in international trade is through the WTO Sanitary and Phytosanitary Agreement (SPS), which uses the IPPC and OIE standards as the base for evaluating SPS disputes. National measures based on international standards from either of these institutions will generally not be open to challenge under the WTO dispute resolution process. Furthermore, both organizations nominate experts for WTO SPS dispute panels and provide technical background information to the panels based on their standards. As such, they can have far-reaching economic and political consequences on food trade.

The *Codex Alimentarius* Commission (Codex), under the joint FAO/WHO Food Standards Program, provides a similar service related to processed foods. Codex develops international food standards, which identify the product and its essential composition and quality factors, identify additives and potential contaminants, set hygiene requirements, provide labelling requirements, and establish the scientific procedures used to sample and analyze the product. Each standard normally takes six or more years to develop. Determination of the safety of the food product is based on scientific risk analysis and toxicological studies. Once a Codex standard is adopted, member countries are encouraged to incorporate it into any relevant domestic rules and legislation but they may unilaterally impose more stringent food safety regulations for consumer protection, provided the different standards are scientifically justifiable. Codex plays an important role in agri-food trade because its standards, guidelines, and recommendations are acknowledged in the SPS and Technical Barriers to Trade Agreements of the WTO. As with the IPPC and OIE standards, during consideration of trade disputes domestic standards based on Codex standards are usually sustained. Turning to new technologies, while there are currently no specific Codex standards in place for products of biotechnology or hormone treated beef, there has been significant effort in Codex to develop a standard for both. (See Chapter 4 for details.)

The OECD, composed of 29 industrial democracies, has actively assisted in the harmonization of international regulatory requirements, standards, and policies related to the discipline of biotechnology since 1985. The OECD has undertaken a number of projects to make regulatory processes more transparent and efficient, to facilitate trade in the products derived through biotechnology, and to provide information exchange and dialogue with non-OECD countries. The OECD leads efforts to develop "consensus documents," which are scientific background documents mutually recognized by member states. These documents set out the biology of the crop plant, introduced trait, or gene product and they provide a common base to be used in regulatory assessment of an agricultural or food product derived through modern biotechnology. By August 2000, 14 consensus documents were published by the OECD (OECD 2000a). In June 1999 the G8 requested the OECD to undertake a study of the implications of biotechnology and other aspects of food safety. This involved, among other actions, an international food safety conference in Edinburgh in February 2000 which concluded that the regulatory systems around the world have to date only approved GM foods that are safe. The OECD reported its findings to the G8 summit in Okinawa in July 2000, which acknowledged the value of the OECD as a forum for continued policy dialogue.

A number of bilateral or multilateral regional initiatives have played, and will increasingly play, an important role in the regulation of trade in goods and services. These institutions play a vital role in creating the consensus necessary to establish international rules, as many food safety concerns are often bilateral, or the knowledge base to develop standards resides in only a few countries. The Trans-Atlantic Economic Partnership (TEP) between the US and EU, for example, has undertaken talks in recent years to improve regulatory processes and scientific cooperation through mutual recognition of testing and approval procedures; progressive realignment or adoption of the same standards, regulatory requirements and procedures; adoption of internationally agreed standards; and dialogue between scientific and other expert advisers in standard-setting bodies and regulatory agencies. The EU has similar trade liberalization initiatives with Canada and

Japan. Since 1998 the Canadian Food Inspection Agency and the United States Department of Agriculture, Animal and Plant Health Inspection Service have also been studying and comparing the molecular genetic characterization of transgenic plants in search of ways to harmonize their regulatory review processes. Some agreement has already been achieved, although no formal binding bilateral agreement has yet been concluded. Meanwhile, the US, Canada, and the EU all offer training and support for regulators in key import markets (usually developing countries) in an effort to "export" their regulatory models to other countries. These bilateral processes could be an important way to resolve technically based trade disputes. Regional agreements, memoranda of understanding, mutual recognition agreements, formal dialogues, and joint research projects are mechanisms that can be used to decrease bilateral regulatory barriers to GM food trade. Using these mechanisms may help countries achieve the greatest trade liberalization that is possible at a given point in time. Then, if a critical mass of regions achieve sufficient levels of liberalization, regional agreements could be relatively easily incorporated into the multilateral international trade regime. In this way regional trade agreements may be complementary, if not preliminary, to the WTO process.

The WTO has become a focal point for examining and resolving trade disruptions resulting from food safety concerns. While there was an agreement on technical barriers to trade in the Tokyo Round of GATT talks, the 1995 agreement, for the first time, extended the newly formalized and binding dispute settlement system to cover trade concerns related to sanitary and phytosanitary rules and technical barriers to trade. SPS and TBT cases accounted for the largest single group of cases brought before the WTO as of July 2000. The WTO has recently ruled on a number of disputes relating to SPS measures and was asked in September 2000 by Thailand to adjudicate a case related to GM foods. While the WTO agreement permits national "standards or regulations for the classification, grading or marketing of commodities in international trade" (article XI), and the adoption or enforcement of measures necessary to protect human, animal or plant life or health (article XX(b)), it sets some rules on when and how they may be used. Specifically, the SPS Agreement specifies that: (i) SPS measures not discriminate between member states; (ii) standards which conform to international standards developed by international organizations (i.e., Codex, OIE, and IPPC) are presumed to be consistent with the obligations outlined in the SPS Agreement; (iii) national standards that are in excess of established international standards or where no international standards exist must be based on scientific principles and the completion of a risk assessment study; and (iv) measures shall not constitute a disguised restriction on international trade.

If a country believes another party has unjustly restricted market access through a SPS measure, it can use the WTO dispute settlement system. Once a case has been presented, heard and adjudicated, the General Council approves the final dispute panel report unless there is unanimous consent not to adopt it. To date, WTO panels have decided three key cases concerning the validity of national SPS measures: the EU hormones case, the Australian salmon case, and the Japanese agricultural products case. In all three cases the contested domestic SPS measure was struck down on the basis that there was no risk assessment completed to support the SPS measure or the risk assessment was improperly done. The salmon case decision sets out that a proper risk assessment must: (i) identify the disease(s) a member wants to prevent within its territory and the potential biological and economic consequences associated with the entry or spread of the disease(s); (ii) evaluate the likelihood of the disease entering and causing damage without the measure; and (iii) evaluate the likelihood of the entry or spread of the diseases with the SPS measures (WTO 1998, para. 121). If the risk assessment does not even refer to the SPS measure, it is doomed to failure before a WTO panel.

While the WTO is a prime target for dispute resolution for many countries, it has some limitations. As currently interpreted, the SPS Agreement allows regulations based on science but does not permit non-science concerns such as consumer preference, animal welfare, or non-measurable environmental risks to be considered in the determination of whether a SPS measure is based on science and supported by a proper risk assessment. WTO dispute cases show that the organization is likely to value unrestricted trade and scientific proof above other factors such as environmental protection or socio-economic considerations. There is a very real risk that WTO decisions that are contrary to domestic concerns may be ignored. This could topple the WTO panel process and the organization as a whole, jeopardizing the other operations of the WTO in trade liberalization.

The BioSafety Protocol is the newest effort to provide a comprehensive international structure to ensure the protection of biodiversity and to facilitate consideration of non-scientific concerns in food trade (see also the chapter by Falkner). While the Cartagena Protocol, concluded in Montreal in January 2000, is primarily designed to provide rules to facilitate advance informed agreement (AIA) for first time transboundary movements of GM organisms intended for environmental release, it also provides for labelling (but not AIA) of GM elements in commodity shipments destined for the food chain. While countries can use this transparency to decide whether to import those commodities, the current interpretation is that import bans will still need to be based on a scientific risk assessment. It is perhaps too early to make a confident evaluation of the protocol. Ultimately, the economic and trade impact of the protocol depends on how it is implemented. Isaac and Phillips (1999) examined the operation of the BSP on the Canadian canola trade and concluded that it could affect between 0.5 percent of total exports (i.e., first-time shipments of GM organisms intended for deliberate release) and 100 percent of shipments (if all shipments, which currently blend GM and non-GM varieties, are judged to be destined for release).

This review of current international institutions does not yield an obvious candidate to resolve the international trade difficulties related to food safety. While traditional concerns about food safety can be and are often adequately handled by the current agencies, none of the agencies is designed to deal with many of the challenges posed by new technologies. Consumer concerns, ethical considerations and various socio-economic factors do not fit within any of the institutions. In the absence of any consensus on those and other issues, national regulatory systems will continue to diverge. As a result, the agri-food community will continue to face a "patchwork" of regulations, with national systems providing basic but different standards and international bodies setting minimum but potentially unenforceable standards for different aspects of the regulation of products of biotechnology. At root, national governments do not have enough confidence in each other to accept their trading partners' systems. For the time being, the agri-food market will continue to face that uncertainty.

## THE PRIVATE SECTOR RESPONSE

Food safety has been become a core requirement for continued operation by most agri-food enterprises. As a result, industry in most OECD member states has moved aggressively to implement new food safety standards and in many cases has invested heavily in the development of private standards for quality and consistency that exceed the regulatory minimums. In short, many firms now self-regulate.

The major effort within industry is focused on developing Hazard Analysis Critical Control Point systems to assure both safety and quality. The original HACCP program, based on the total quality management theories of W. Edwards Deming, was developed in the 1960s by the Pillsbury Company, the US Army, and NASA as a collaborative effort to produce safe foods with "zero defects" for the NASA space program. Pillsbury replaced traditional quality systems based on finished product sampling and testing with a new process that used operator control and/or continuous monitoring techniques at critical control points. Pillsbury extended that program into its commercial food production systems during the 1960s and publicized its form of the HACCP concept publicly at a conference for food protection in 1971. By 1974 the US Food and Drug Administration had adopted HACCP principles in new low-acid canned food regulations. In 1991

Codex initiated a working group to formalize a worldwide approach and application of HACCP principles. Most of the major exporting countries in the OECD indicate that they are pursuing HACCP-style risk management in their agri-food systems. Canada's Food Safety Enhancement Program is one example of an HACCP model being incorporated into the regulatory system (Codex 2000).

The HACCP model has become the quality assurance program of choice for many food producers and exporters. When the international regulatory system evolves too slowly to meet industry needs, industry has in the past moved to models for self-regulation and coordination. Thus, one possible outcome of slow development of international regulations might be for the companies or parts of the industry (e.g., the biotechnology industry) to implement self-regulation to maintain market access. There are a number of cases where parts of the agri-food industry have developed systems to deliver products with higher standards than domestic or even international minimum standards. The red meats industry in Australia (Spriggs and Isaac 2001), the canola industry in Canada (Gray, Malla and Phillips 1999; Smyth and Phillips forthcoming), and retailers and processors in the EU, North America, and Asia (Phillips and McNeill forthcoming) have all adopted private standards at one time or another in recent years.

This is especially relevant for the biotechnology industry. The introduction of GM foods into the agri-food market has split it into two distinct markets and to date there has only been limited interaction between the two. Some markets, such as the EU, and some food processors, have decided to forgo GMOs for now, and are devoting increasing effort to securing adequate volumes of GM-free foodstuffs to satisfy their customers. Consumers in those markets for the most part do not have any opportunity to consume GM foods, as they simply are not available. Other markets, such as in North America and in many product lines (maize products), have rapidly adopted the technology and for the most part do not offer a choice of GM-free food to their consumers. Trade between these two blocks has slowed dramatically in all those product markets where GM varieties are being used. In response, some firms have begun to invest to develop quality assured supply chains to be able to bridge the gap between those two markets.

While private labelling systems now abound, actually delivering the promised product has proven more difficult. The main difficulty of buying from markets with both GM and non-GM varieties is that, as of September 2001, tests for determining whether products are GM-free are costly, often inaccurate and always slow. As a result, companies are being forced to develop identity preserved production and marketing systems to segregate product lines, to control the flow and to assure the quality of GM-free produce to serve those markets. These systems are costly and markets, at least in early 2000, did not appear to offer a large enough premium to compensate for the costs (Lin 2000). There are a number of examples in Canada where producers in the canola industry have for limited periods adopted identity-preserved production and marketing systems for GM and non-GM novel trait varieties, but the resulting systems have imposed incremental costs equal to 10 to 15 percent of farm gate prices (Phillips and Smyth 2000; Smyth and Phillips forthcoming).

## CONCLUSIONS AND OBSERVATIONS ON THE WAY AHEAD

World food markets are faced with a conundrum. Science has advanced more rapidly than either regulators or the markets are able to manage. International institutions, in particular, have not been able to find international consensus on standards that can form a generally accepted base for trade in food. As a result, there has been a *de facto* re-nationalization and in some cases privatization of food safety regulation. Given the significant international trade in foodstuffs, this has led to trade disputes and distorted trade flows, as national governments and private supply chains have responded to citizen and consumer concerns. Compounding this, there is serious doubt that regulators in every country now have or can sustain the necessary technical capacity to adequately assess the risks of the rapidly evolving array of products entering the market.

This raises three basic questions. First, does it matter? Second, is there a better alternative? Third, how can we move toward that alternative? This chapter offers tentative answers to all three questions.

First, this rising level of conflict does matter. There is real concern that the current international institutions may not be able to manage the conflicts, which could work to destabilize the entire postwar trading system for all products and services. We all would lose, especially consumers. Few nations can truly claim to be self-sufficient in foodstuffs. Most raw or semi-finished commodities and key agricultural inputs (energy, fertilizer, chemicals) depend on a functioning trade system. Even if a nation is a net food exporter (as is Canada), there is a strong likelihood that it either cannot produce the desired variety of foods (due to climate) or cannot produce a steady supply of foods (due to seasons). So we have a mutual dependence in the agri-food system. Even if we muddle along, there is a real concern that the pressures for change will rise. As new GM varieties are introduced into new product lines and new markets, public concern is rising and domestic regulatory oversight of production and trade is increasing. As time passes, the yield-enhancing effect of new technologies will expand production in key export markets, further depress world prices and increase demand for freer flow of GM food products. In the absence of any international consensus on how to jointly regulate the safety of these foods, national regulators will continue to enact rules that will effectively reduce international trade, which if sustained long enough will dampen private research, development, and investment in the agri-food sector, which will hurt all consumers.

Second, there is ample evidence from theory and history to show that international consensus on food safety issues is both possible and desirable. The simple conclusion is that the scale of diffusion of new technologies is so large that governments throughout the world must consider how to more effectively regulate production and trade. Most of the domestic and international efforts currently underway are trying to find comprehensive solutions to all the concerns that are being raised about both current and all future food safety risks resulting from new technologies. While that is laudable, it may not be possible. Instead, perhaps we need to find the answer in the past, when specific solutions were found to actual specific problems. In this context, perhaps it would be possible to look initially for a patchwork of inconsistent but functioning standards for these new technologies in these new products. So, for instance, we might have one standard for canola, where only the oil is consumed by humans, and another for maize and soybeans, where both the oil and meal are consumed by humans. Similarly we might have differing standards for products that have intra-species genetic transformations compared with those with transformations involving only one species. In short, we may be able to find practical solutions to current, real problems. Related to that, we may need to consider adopting greater specialization in the regulatory system. There is some evidence that the intellectual capital and research capacity for many products is concentrated in one or a few research centres (Phillips and Khachatourians 2001). If that trend holds, and regulators need access to those resources (which many argue they do), then conceivably it may be prudent to have the regulatory function undertaken by the nation (or nations) with research capacity in those products. In short, perhaps we should consider a future with specialized research, regulation, and production integrated in the key producing nations. This would ultimately involve many sites of authority in the international system, including national regulators, supranational agencies and private organizations. It will be necessary to look closely at how these different groups would interact and operate. Currently, many of them operate competitively but there may be benefits in having them coordinate or collaborate more.

Third, a portfolio of approaches will be needed to comprehensively manage food safety. This will likely include continued science-based consensus-building through institutions (such as the IPPC, OIE, and Codex), bilateral or multilateral procedural consensus-building through regional efforts (such as the TEP or at the OECD), multilateral negotiations and precedent-setting through dispute settlements (at the WTO and the BSP), and capacity-building through the Consultative Group on International Agricultural Research. Perhaps we will also need to modify our methods of negotiating. Recently most international negotiations have involved a large number of interests and tended to seek formula solutions to a wide range of discrete problems. That may work for negotiating changes to international tariffs and quotas, where there is consensus about the relative size and impact of the measures and they can be quantified according to generally accepted rules. It offers little help, however, to those trying to negotiate trade rules related to food safety measures where the fundamental problem is a lack of agreement about the goals and objectives. Somehow, we need to find a way to build consensus. One possible approach would be to return to the days of old, where negotiations tended to be "bid and ask" sessions focused on specific products or markets between the two nations with the most at stake. This process distilled the debate down to those elements where consensus was possible and then "multilateralized" the results through the most-favoured-nation provision. A modified process involving key actors could help to break the log-jam. While this is possible, it would not likely be easy. In this case, the two areas with most at stake are the US and the EU, which have adopted policies that reflect different core values. Hence, even though they have the most at stake, they do not necessarily have any means to negotiate their way out of the impasse. Furthermore, even if they could reach some compromise, that would not necessarily provide a basis for a global settlement. Many other developed and all of the developing countries have other concerns that are unlikely to be addressed in such a bilateral process.

Even if all the above were to occur, this would still leave a noticeable gap at the international level for those individuals, groups, and states that focus on food in terms of its socio-economic dimensions. Some have suggested that a new body or international forum, like the Intergovernmental Panel on Climate Change, is needed to handle the specific issue of the socio-economic implications of biotechnology (Krebs 2000). The fundamental problem is that while there are many socio-economic issues related to biotechnology, many (most) of them are not unique to biotechnology or even to food. Questions as to the rights of development, international and intergenerational equity, indigenous people's or farmers' rights, the impact of international agreements on farmers and consumers and even ethical questions concerning the exploitation of life on the planet are not unique to food. Biotechnology and food may simply be a proxy issue for these larger problems rather than the real issue to be resolved. It might be better, then, to engage existing international bodies such as those within the human rights system and those dealing with development and the environment to establish a coordinating committee to examine broad-based, socio-economic concerns including those arising from biotechnology. If non-scientific issues have no credible means of being heard and considered, some groups may believe that the only means to redress their concerns is to derail any international trade or regulatory efforts. Although some preliminary steps have been taken to address socio-economic issues arising from biotechnology (the BioSafety Protocol, for example), more research and government-to-government dialogue is required to craft a process for the expression and evaluation of socio-economic concerns in the modern economy.

In conclusion, the new technologies raise radical new issues for regulators, industry, consumers, and citizens. Perhaps it is time to consider radical solutions in industry and government to address these concerns.

## REFERENCES

Angus Reid Group Inc. and *The Economist*. 1999. *World Poll*. Angus Reid Group Inc., London, <http://www.ccu-cuc.ca/en/polls/data/angus.html>.

Buckingham, D., R. Gray, P. Phillips, T. Roberts, J. Bryce, B. Morris, D. Stovin, G. Isaac and B. Anderson. 1999. "The International Co-ordination of Regulatory Approaches to Products of Biotechnology." Ottawa: Agriculture and Agri-Food Canada. <http://www.ag.usask.ca/departments/agec/cmtc/marketing.htm>.

*Codex Alimentarius* Commission (Codex). 2000. "Origins of the *Codex Alimentarius*." 12 October. <http://www.fao.org/docrep/w9114e/W9114e03.htm>.

Doern, B. 2000. "Inside the Canadian Biotechnology Regulatory System: A Closer Exploratory Look." Ottawa: Canadian Biotechnology Advisory Committee, November. <http://www.cbac.gc.ca/english/reports/listdocs.aro?i=CBAC+Commissioned+Reports&type=42>.

Einsedel, E. 2000. "Meeting the Public's need for Information on Biotechnology." Ottawa: Canadian Biotechnology Advisory Committee, November. <http://www.cbac.gc.ca/english/reports/listdocs.aro?i=CBAC+Commissioned+Reports&type=42>.

Food and Agriculture Organization (FAO). 2000. Data Collections. <http://apps.fao.org/cgi-bin/nph-db.pl?subset= agriculture>.

Gray, R., S. Malla and P.W.B. Phillips. 1999. "The Public and Not-For-Profit Sectors in a Biotechnology-Based, Privatizing World: The Canola Case." Proceedings of the NE-165 Conference, "Transitions in Agbiotech: Economics of Strategy and Policy," Washington, DC, 24-25 June.

Isaac, G. and P.W.B. Phillips. 1999. "Market Access and Market Acceptance for Agricultural Biotechnology Products." Proceedings of the ICABR conference on "The Shape of the Coming, Agricultural Biotechnology Transformation: Strategic Investment and Policy Approaches from an Economic Perspective," University of Rome "Tor Vergata," 17-19 June.

James, C. 2000. "Preview: Global Review of Commercialized Transgenic Crops: 2000," ISAAA Brief No. 21-2000. <http://www.isaaa.org/publications/briefs/Brief_21.htm>.

Krebs, J. 2000. "GM Food Safety: Facts, Uncertainties, and Assessment." Final report of the OECD Edinburgh conference on the scientific and health aspects of genetically modified foods. <http://www.oecd.org/subject/biotech/edinburgh.htm>.

Lin, W. 2000. "Biotechnology: Grain Handlers Look Ahead." <http://usda.mannlib.cornell.edu/reports/erssor/economics/ao-bb/2000/ao270.as>.

McHuhen, A. 2000. *Pandora's Picnic Basket: The Potential and Hazards of Genetically Modified Foods*. Oxford: Oxford University Press.

Organisation for Economic Co-operation and Development (OECD). 2000*a*. "Regulatory Developments in Biotechnology in OECD Member Countries." <http://www.oecd.org/ehs/country.htm>.

_____ 2000*b*. "Biotrack Online: Consensus Documents." <http://www.oecd.org/ehs/cd.htm>.

_____ 2000*c*. "Compendium of National Food Safety Systems and Activities." 7 June. <SG/ADHOC/FS(2000)5/ANN/FINAL>.

_____ 2000*d*. "Overview and Compendium of International Organisations with Food Safety Activities." 11 May. <SG/ADHOC/FS(2000)4/FINAL>.

Phillips, P.W.B. 2001. "International Trade in Genetically Modified Agri-Food Products," in *Agricultural Globalization, Trade and the Environment*, ed. C. Moss, G. Rausser, A. Schmitz, S. Taylor and D. Zilberman. Boston: Kluwer.

Phillips, P.W.B. and D. Buckingham. 2001. "Agricultural Biotechnology, the Environment and International Trade Regulation: Your Place or Mine?" in *Globalization and New Agricultural Trade Rules in the 21ˢᵗ Century*, ed. H. Michelmann, J. Rude, J. Stabler and G. Story. Boulder, CO: Lynne Rienner.

Phillips, P.W.B. and G. Khachatourians, eds. 2001. *The Biotechnology Revolution and Global Agriculture: Innovation, Invention and Investment in the Canola Industry*. Oxford: CABI Pub.

Phillips, P.W.B. and H. McNeill. 2001. "Labeling for GM Foods: Theory and Practice," *AgBioForum* 3(4).

_____ Forthcoming. "Labeling for GM Foods: Theory and Practice," in *Market Development for Genetically Modified Foods*, ed. V. Santaniello, R.E. Evenson and D. Zilberman. New York: CABI Pub.

Phillips, P.W.B. and S. Smyth. 2000. "Managing the Value of New-Trait Varieties in the Canola Supply Chain in Canada." Proceedings of the 4ᵗʰ Wageningen Supply Chain Conference, May.

Smyth, S. and P. Phillips. Forthcoming. "Competitors Co-operating: Establishing a Supply Chain to Manage Genetically Modified Canola." Special issue of the *International Food and Agribusiness Management Review*.

Spriggs, J. and G. Isaac. 2001. *Food Safety and International Competitiveness: The Case of Beef*. Oxford: CABI Pub.

Strategis. 2000. Data search of Canadian exports by product and market. Industry Canada. <http://strategis.ic.gc.ca/engdoc/main.html>.

World Trade Organization (WTO). 1998. Appellate Decision. *Australia — Measures Affecting Importation of Salmon — Complaint by Canada*. AB-1998-5. (Appeal from WT/DS18.) <http://wto.org>.

# 4

# International Efforts to Label Food Derived through Biotechnology

*Anne A. MacKenzie*

Genetically modified (GM) plants, and food products from biotechnology,[1] have recently been introduced into the market. They are generating significant consumer concerns and have led a number of governments to adopt new rules regarding the labelling of GM products. While many of these national labelling schemes are designed to address local consumer concerns, their inconsistent and incomplete nature exacerbates uncertainty, as consumers face greater difficulty in understanding the nature of their purchases, and further complicates the operation of the international food trading system.

Many countries have observed the potential benefits presented by biotechnology and have drawn up strategic plans and regulations to guide its exploitation for national needs. In addition, a large number of international organizations have addressed, or are in the process of addressing, food biotechnology. To date, much of the international activity has focused on an attempt to develop appropriate biotechnology policies for labelling GM foods, as well as the establishment of guidelines for harmonization. One of the key organizations in this field is the *Codex Alimentarius* Commission (Codex).

The potential for labelling to enhance consumer choice is obvious. Also obvious is the potential for GM labelling to be used for protectionist reasons. A label suggesting that GM products are unsafe could result in the limitation of imports. However, even a label that is truthful can be misleading. Unless definitions are clear, consumers can be misled; a label that requires a standard not achievable in nature, or that requires an impossible degree of segregation, can be used as a disguised barrier to imports.

This chapter provides an overview of some of the key concerns surrounding GM food products and details the procedures followed by Codex in its role as a forum for the development of international standards. The ongoing debate with respect to the development of an international labelling regime for biotechnology-derived food products illustrates the difficulties associated with Codex's desire to arrive at an international consensus.

## LABELLING GENETICALLY MODIFIED FOODS

One of the most challenging issues surrounding GM foods today involves labelling. Genetically modified organisms (GMOs), in regard to plants used for food, have become a high concern due to differences in labelling requirements in various countries. Both nationally and internationally, food labelling issues are becoming more complex, as new issues continue to arise regarding safety, social, cultural, environmental, and trade concerns. There is considerable debate at the international level about the labelling of biotechnology-derived foods.

On the one hand, it is argued that foods derived through biotechnology are not fundamentally different from conventional or traditionally manufactured foods, and as such, must follow the same labelling requirements as those applied to conventional foods. This would include voluntary claims about the method of production where appropriate. On the other hand, consumers argue that they have a "right-to-know" each product's method of production. Mandatory method of production labelling thus becomes necessary in order to enable individuals to make informed decisions about the foods they consume.

Critics cite three potential types of health risks for consumers which may be caused by unintended effects of biotechnology, and for which all GMOs must require labelling. First, there is the risk that food allergens will be transmitted to the host organism. Second, food obtained through biotechnology could differ significantly from the corresponding food with regard to composition. Third, modifications to foods could alter their nutritional content. There may also be the possibility of an environmental impact as a result of gene escape, which could create herbicide resistance in weedy relatives of the transgenic herbicide crop.

Concern about such potential health risks is increasing in Europe and Japan. The European Union (EU) and Japan have created regimes to regulate the processes used to create novel plants and plant products. Consumer concerns with regard to the impact of genetic modifications have driven the EU to require that all food and feed items that contain DNA protein, and have undergone significant compositional changes as a result of biotechnology, be labelled as containing GMO material.

For the North American consumer, regulators in Canada and the United States propose that a food product be labelled for its new characteristics only if it has been changed in a significant way or presents a safety concern to certain segments of the population. For example, food labels would indicate the introduction of an allergen into the food as a result of genetic engineering. In this case, food labelling would allow consumers to make informed product choices about nutrition, composition, and substances that may have safety implications.

As the labelling debate continues, countries will be weighing the pros and cons associated with the mandatory labelling of genetically engineered food. Several issues have emerged over the past years regarding the threshold for labelling a food product as a GM product. Do different policies and regimes for regulating biotechnology-derived foods around the world impede international trade? How much information must be required on a label? Do consumers have a "right-to-know?" Should the labelling of biotechnology-derived foods be mandatory, voluntary, or a hybrid of both? Is there a precise way to detect every GMO protein? Does labelling products with phrases such as "GMO-free" or "non-GMO" connote a claim to product superiority? These are just a few of the issues that are driving the current debate regarding the labelling of biotechnology-derived foods.

## FOOD SAFETY

The health effects of GM foods is dependent on the specific content of the food itself, and may have either potentially beneficial, or occasionally harmful health effects if the process involves the introduction of an allergen. Allergenicity is an important consideration for foods derived through biotechnology because of the possibility that a new protein introduced into a food could be an allergen. This concern is particularly relevant for genes that are derived from foods that commonly cause serious, life-threatening allergic reactions. One has to carefully consider the possibility of evaluating and identifying such potential risks prior to commercialization. This task is normally performed by regulatory agencies prior to a product's approval.

In Canada, new products are reviewed against a set of rigid standards established by Health Canada. Before any genetically modified micro-organism or plant is approved by Health Canada as a safe food, it must undergo a thorough assessment. If there exists any health or safety concern related to allergens, the department requires clear labelling. If the assessment demonstrates that the risks cannot be mitigated through labelling, then the food will not be approved. In Canada, foods derived through biotechnology require labelling under three conditions:

- when any food, or food ingredient, obtained through modern biotechnology contains an allergen transferred from any of the products causing hypersensitivity, the allergens shall be declared;

- when products obtained through biotechnology differ significantly from the corresponding food with regard to composition; or

- when there is a significant difference from the corresponding food with regard to its nutritional value and intended use.

In general, with regard to health issues, tests on toxicity and allergenicity have been, and continue to be, conducted. To date, none have shown significant toxic or allergenic harm. No peer-reviewed article or epidemiological study reporting adverse effects on human health has yet appeared (OECD 2000a). Yet, the current uncertainty about whether genetic modification could lead to the transfer of unknown allergens, demonstrates the need for regulators to be alert and to continually refine protocols.

## CONSUMER INFORMATION

In general, food labelling enables consumers to make informed product choices with respect to nutrition, composition, and substances that may have safety implications for particular segments of the population. Consumers' increased knowledge about diet and health; concern about food safety and misrepresentation; and access to information about new production and processing technologies, have increased the pressure for greater label information. However, the issues associated with modern biotechnology go beyond information about product characteristics. It has been maintained that the right of consumers to make informed choices should be respected, and that reliable labelling is the only means to ensure consumer confidence in this area, even if this means broadening the basis for labelling requirements.

Consumer education with respect to food biotechnology is of crucial importance in order to ensure its acceptance. To be of use to consumers, label information must be communicated in a manner that is not misleading. For example, the appearance of the label, and the language used on the label, should not imply that the consumption of a food derived through biotechnology has implications for public health, unless there is enough scientific data available to support this implication. Conversely, if a food is labelled as non-GMO or GMO-free, the label should not imply that the food is healthier than products that may contain ingredients of biotechnology.

## DETECTION METHODS AND TRACEABILITY

As the debate over the safety and labelling aspects of GM foods is intensifying, so are the issues of testing and traceability. Consumer demands for labels on GM foods require a precise method of detecting and measuring all the chemical signatures left in products by the processes of genetic modification. Their demands also require validated analytical methods to establish "thresholds" below which labelling will not be required. For example, commodity products such as corn and soybeans are generally handled in bulk, meaning that different crop varieties of a product are mixed during harvest to achieve a certain grade.

Due to the nature of food production, marketing, and handling, it would be very difficult to guarantee that a product does not contain any amount of biotechnology-derived product. Even the use of the same processing equipment to handle food products could result in the presence of small amounts of biotech material in a conventional crop shipment. Thus, workable policies for labelling should take into account threshold values. Standards could be established to govern the amount of biotech material present in a sample that would require the product to be labelled as containing biotech ingredients.

Presently, there is no single test that can be used to detect all GM products and plants. Furthermore, there are no international standards, or established thresholds, for GM food testing and sampling. Three main categories of tests for the detection of GM products and plants currently exist (Lanterman 2000); however, an important consideration in using such tests is that some methods are proprietary and owned by private companies. Direct DNA-based tests probe for the presence of a specific inserted genetic sequence. These tests include methods for the detection of specific DNA sequences (e.g., Polymerase Chain Reaction method), and, non-specific DNA sequences, which are present in most GM plants. Indirect protein-based tests search for the presence of the protein expressed by the specific genetic sequence. Indirect protein-based tests, such as the ELISA test, tend to be simpler than the direct tests, but their weakness is that they cannot identify the specific genetic modification. Bioassays test for the production of a particular trait in viable seeds, such as resistance to a herbicide.

It should also be noted that the effectiveness of most methods are not internationally validated or accepted. International validation involves determining the sensitivity and specificity of the test (its ability to identify accurately all positive and negative cases). Currently, in theory at least, screening methods, such as testing for flanking gene sequences, permit the detection of many different GMO events and are used more commonly than specific detection methods.

## TRADE RULES

When the Sanitary and Phytosanitary (SPS) and Technical Barrier to Trade (TBT) Agreements were negotiated, GMOs were not a trade issue. How then might the rights and obligations of the SPS Agreement relate to trade in GMOs? One of the first things to consider is whether there are international standards applicable to GMOs. Currently, there are no international standards that specifically govern these, nor is there a harmonization of regulatory approaches mandated. However, the SPS and TBT Agreements have spurred countries to modify their regulatory systems.

The SPS Agreement recognizes the rights of countries to establish their own levels of protection, and to impose measures necessary to protect human, animal, and plant health. The agreement relies largely on the principle of scientific evidence in order to avoid arbitrary discrimination against imports. In doing so, the agreement gives special status to international standards. If a country bases its measures on the applicable international standards, those measures are presumed to be in compliance with the SPS Agreement. Countries are not obligated to adopt international standards. A state may choose to impose a measure that is not based on an international standard, even if it provides a higher level of protection, provided it does so on the basis of a scientific risk assessment.

Consumer attitudes toward risk and government approaches to food safety and quality vary significantly from country to country. Differences in attitude and regulatory stances are potential contributors to trade disputes. National and international regulation of products of modern biotechnology, in particular the segmentation of some markets into GMO and non-GMO products, will have a direct impact on international trade.

The TBT Agreement permits governments to impose labelling requirements that are necessary to fulfill legitimate objectives. However, members disagree on whether providing information to consumers on the method of production that is unrelated to end-product characteristics is a legitimate objective. Given the growing debate on biotechnology, food safety and labelling issues are likely to remain a priority on the trade agenda.

## THE CODEX COMMITTEE ON FOOD LABELLING

Codex was created in 1962 to implement the joint Food and Agriculture Organization (FAO) of the United Nations/World Health Organization (WHO) Food Standards Program. Membership in Codex is open to all member nations of the UN. Currently, 165 countries participate. The Food Standards Program, as demonstrated by more than 4,000 standards, recommendations, and guidelines accepted to date, strives to protect consumer health and ensure fair trade practices involving food. The program involves the determination of priorities and provides guidance for the preparation and finalization of standards. These standards are referred to in the WTO Sanitary and Phytosanitary Agreement and are published either as regional, or worldwide, standards.

Once a Codex standard has been adopted, member countries are encouraged to incorporate it into all relevant domestic rules and legislation. However, under the SPS Agreement, member countries retain the right to unilaterally impose more stringent food safety regulations, if they are deemed necessary to ensure domestic consumer protection. Heightened food safety regulations, above and beyond Codex standards, must be scientifically justifiable and otherwise consistent with SPS rules. The Codex Committee on Food Labelling (CCFL) is considering the issue of GM food labelling in an effort to develop guidelines for international harmonization.

The *Procedural Manual* for Codex outlines a uniform procedure for the elaboration of Codex standards. In brief, there are eight key steps to developing a new standard. First, the commission, or a subsidiary body, decides to elaborate a worldwide Codex standard and identifies which subsidiary body, or other bodies, should undertake the work. In the case of Codex regional standards, the commission bases its decision on the submitted proposal that is supported by the majority of members belonging to a given region, or group of countries. Second, the secretariat arranges for the preparation of a proposed draft standard, drawing on expertise from the FAO, WHO or the International Dairy Federation. The third step involves sending the draft standard to members of the commission and interested international organizations for comment on all aspects, including possible economic implications of the proposed standard. Fourth, the received comments are sent by the secretariat to the subsidiary body or other body concerned, which has the power to consider such comments and to amend the proposed standard. Fifth, the secretariat submits the draft standard to the commission or to the executive committee, which considers all input from members or the subsidiary body as it discusses whether to adopt the draft standard. Sixth, the revised draft standard is once again sent to all members and interested international organizations for comment. During the seventh stage, the comments received are sent by the secretariat to the subsidiary body, or other bodies, for consideration and further amendment of the proposed standard. Finally, the draft standard is submitted through the secretariat to the commission, together with any written proposals received from members and interested international organizations for amendments, and a decision is made whether to reject or adopt the measure as a Codex standard.

In the case of regional standards, all members and interested international organizations may present their comments, take part in the debate and propose amendments, but only the majority of members of the region, or group of countries concerned, in attendance at the session can decide to amend and adopt the draft. This procedure normally takes more than six years to complete. There is an accelerated procedure that shrinks the process to the first five steps, but a two-thirds majority is required to proceed in this manner.

Hosted and chaired by Canada, the CCFL examines international food-labelling issues, drafts labelling provisions that are applicable to all foods, and endorses labelling provisions prepared by Codex committees charged with drafting standards, codes of practice, and guidelines. The high degree of controversy generated by many issues before the CCFL, demonstrates the significance that member countries accord to the development of international labelling standards. The CCFL considers the major issues around the labelling of biotechnology-derived foods. To date, the CCFL has advanced section 2 (Definition of Terms) and section 5 (Additional Mandatory Requirements—allergens) to Step 6 of the Codex Acceptance Procedure.

The following summary of the CCFL's seven-year effort in developing a standard for the labelling of biotechnology-derived foods demonstrates the laborious process involved in developing a Codex standard. The objective is to develop a standard that reflects input from all governments. Each of the meetings discussed below were held in Ottawa, and were attended by representatives of some of the Codex member countries, international consumer groups, private industries, and by representatives from the Codex secretariat and observers.

*22nd Session (April 1993).* The CCFL agreed that work on the labelling aspects of biotechnology be considered in light of recommendations by the Codex commission. The CCFL requested the US to prepare a discussion paper for consideration at the next session.

*23rd Session (October 1994).* The committee considered the discussion paper prepared by the US. The paper identified a number of issues as areas where further elaboration and comments should be sought. During these initial discussions, countries either favoured mandatory labelling only for the introduction of any potential health or safety concerns to food products, or advocated that labelling be required under all circumstances. Some countries thought that it was too early to determine particular rules for products obtained through biotechnology. They felt that labelling should be required only when the food or ingredient differed significantly from its traditional equivalent or if safety concerns were involved, such as the introduction of an allergen.

*24th Session (May 1996).* Delegations and observers requested that all food products prepared with the assistance of biotechnology be subjected to mandatory comprehensive labelling. They reasoned that consumers should be able to make choices based on several considerations, including food origin, production method, agronomic practices, and personal values. Some observers also suggested that the public be notified, through labelling, of specific concerns relative to safety, nutrition, and food composition. It was further suggested that these concerns be the subject of scientific evaluation. The EU stated that taking a position on such matters would be premature, as member countries were still reviewing their respective situations. Canada indicated that its policy regarding the labelling of biotechnology-derived foods was still being developed. Noting the lack of consensus, the CCFL agreed to seek guidance from the Codex executive committee on how labelling guidelines might be established.

*25th Session (April 1997).* A draft guidelines document, based on recommendations by the Codex executive committee, was introduced for discussion. Within the *Recommendations for the Labelling of Food Obtained through Biotechnology* (27 February 1997), the executive committee proposed that foods that are not equivalent to existing non-biotech foods with respect to composition, nutritional value, or intended use, should be labelled. The document also suggested approaches for addressing allergens. In order to identify issues and provide direction to the Codex executive committee, the CCFL agreed to solicit comments from Codex member governments. A review comprising these comments was released in February 1998.

*26th Session (May1998).* The draft guidelines document was again discussed. The proposal for labelling foods that are non-equivalent to existing foods, based on composition, nutritional value, or intended use, remained intact. This session provided an opportunity for Codex members to comment on whether all genetically modified foods, or foods that contain genetically modified material, should be labelled. The CCFL facilitated constructive discussion among Codex members. This time, progress was made in refining the definition of products obtained through biotechnology and on the mandatory labelling of foods with allergens, with the exception of food products that are non-equivalent compositionally, nutritionally, or in their intended use. Several European countries, along with India, expressed a preference for the mandatory method of production labelling for all biotechnology-derived foods. Canada, the United States, Australia, New Zealand, Peru, and Brazil supported the labelling of foods based on safety, composition, intended use, and nutrition, which was consistent with their respective labelling laws. The CCFL agreed to forward to the commission for adoption at step five, the definitions related to biotechnology and the provisions on allergens, and to return all other sections of the proposed draft for further consideration.

*27th Session (April 1999).* The CCFL considered a rewrite of the *proposed draft recommendations* (based on the draft guidelines document) and established an ad hoc working group for this purpose. Canada was selected to coordinate and chair the group which comprised representatives from 23 member countries, the EU, and nine international non-governmental organizations. The committee also recommended that a smaller drafting group (consisting of Japan, Brazil, the US, Australia, Canada, and two representatives from the EU) be formed within the working group to "hold the pen." The approach was that the drafting group would write the document and circulate it to the working group for review and comment. The final draft of the recommendations were to be discussed at the CCFL meeting in May 2000. The drafting group reviewed and revised the texts for the definition of biotechnology-derived foods, and the two labelling options being considered by the CCFL.

The mandate of the working group was to develop more fully the two options. The first option requires labelling when products obtained through biotechnology differ significantly from the corresponding food with regard to composition, nutritional value, or intended use. The second option requires the declaration of the method of production for foods and ingredients composed of, or containing genetically modified/engineered organisms, or food, or food ingredients, produced from but not containing GMOs, if they contain protein or DNA resulting from gene technology, or differ significantly from the corresponding food.

The working group also agreed that consideration should be given to the establishment of a threshold level in food and food ingredients, for the presence of food or food ingredients obtained through modern biotechnology, below which labelling would not be required. Consideration would also be given to establishing a minimum threshold level for adventitious or accidental inclusion in food or food ingredients, of food or food ingredients obtained through biotechnology.

*28th Session (May 2000)*. Recognizing the diversity of opinions among member countries, the CCFL engaged in lengthy debate and decided to return the proposed draft for further consideration. The CCFL also agreed that the working group would continue its deliberations to combine options 1 and 2 into a Codex *Guideline*, in light of the proposal from member countries, and to circulate it for consideration by the next session. The working group was also asked to table a paper on key issues and questions associated with the labelling of these foods. A draft discussion document of this nature has since been developed by the US for discussion. Three new members (South Africa, Thailand, and India) were added to the drafting group, which met in India in late October 2000. At that meeting some additional options were developed for consideration during the May 2001 29th Plenary Session.

*29th Session (May 2001)*. Consistency regarding the definition of terms became the major topic of debate when the committee met in Ottawa. Two substantive matters were considered with respect to food biotechnology labelling. In the first instance, the committee agreed to use the definition of "modern biotechnology" adopted by the Cartagena Protocol and moved the definitions to stage eight for decision, which is a critical step. The committee, however, was not able to proceed further with its consideration of the *Guidelines for the Labelling of Foods and Food Ingredients Obtained through Certain Techniques of Genetic Modification/Genetic Engineering*, and returned the current text to step three for further comments. The following terms were agreed upon:

- *Food and food ingredients obtained through certain techniques of genetic modification/genetic engineering* means food and food ingredients composed of or containing genetically modified/engineered organisms obtained through modern biotechnology, or food and food ingredients produced from, but not containing genetically modified/engineered organisms obtained through modern biotechnology.

- *Organism* means any biological entity capable of replication, reproduction or of transferring genetic material.

- *Genetically modified/engineered organism* means an organism in which the genetic material has been changed through modern biotechnology in a way that does not occur naturally by multiplication and/or natural recombination.

- *Modern biotechnology* means the application of

  1. in vitro nucleic acid techniques,[2] including recombinant deoxyribonucleic acid (DNA) and the direct injection of nucleic acid into cells or organelles, or

  2. fusion of cells[3] beyond the taxanomic family, which overcome natural physiological, reproductive or recombination barriers and which are not techniques used in traditional breeding and selection.

As the above discussion has demonstrated, the procedures for the development of international standards are complex, time-consuming, and fraught with difficulty. Nevertheless, the Codex process that began in April 1993 is progressing, albeit incrementally, toward the elaboration of an international standard governing the labelling of biotechnology-derived food and food products.

## CONCLUSION

Agricultural biotechnology clearly has great potential to help solve agricultural production and food security challenges. But the practical impact of food biotechnology in the new millennium is difficult to project. The scientific knowledge that is critical to the advancement of biotechnology is expanding quickly. Recognizing the diversity of consumer perception from country to country, the policy arena is unsettled and is struggling to come up with effective policies to address the concerns and issues.

Issues of major concern include the evaluation of any risks to human health and the environment, the need for mandatory or voluntary labelling of GM foods for domestic consumption and international trade purposes, and the relationships between countries' responsibilities under the WTO and international environmental treaties.

Canada, the United States, Mexico, Japan, and the European Union have considered the appropriate role of labels in signalling new production methods to consumers. Each government currently regulates the introduction of GMO products, but only the EU and Japan require labels that specify the presence of GMOs. As there are no universal definitions of the risks GMOs may present, countries may, and often do, disagree on the type and necessity of information to be provided.

The GM labelling issue is a good example of the complexity of the Codex process. It also demonstrates how narrow technical issues, once only of interest to specialists, have become public policy issues of huge economic importance, imbued with differing social, cultural, and political values. The establishment of standards and the resolution of disputes become particularly difficult where science is relevant but not determinative, and where an international standard will clearly create economic winners and losers.

The current debate within Codex concerning the labelling of foods derived from biotechnology is indicative of how biotechnology is perceived from country to country. The Codex process for standards development is based on reaching international consensus. Based on its past accomplishments in developing labelling standards for other food products, it is hopeful that the CCFL will continue to meet its objective: to protect the consumer and facilitate trade by developing the best labelling policies for harmonization.

NOTES

1.  For the purpose of this document, "food and food ingredients obtained through certain technologies of genetic modification/genetic engineering" means food and food ingredients composed of or containing genetically modified/engineered organisms obtained through modern biotechnology, or food and food ingredients produced from, but not containing, genetically modified/engineered organisms obtained through modern biotechnology (ALINORM 01/22, Appendix V) [9].
2.  These include but are not limited to: recombinant DNA techniques that use vector systems and techniques involving the direct introduction into the organisms of hereditary materials prepared outside the organisms such as micro-injection, macro-injection, chemoporation, electroporation, micro-encapsulation, and liposome fusion.

3. Fusion of cells (including protoplast fusion) or hybridization techniques that overcome natural physiological, reproductive, or recombination barriers, where the donor cells/protoplasts do not fall within the same taxanomic family.

## REFERENCES

*Codex Alimentarius* Commission (Codex). 1993–2000. *Reports of the Twenty-Second Through Twenty-Eighth Sessions of the Codex Committee on Food Labeling.* Ottawa: *Codex Alimentarius* Commission.

Economic Research Service. 1999. *Value-Enhanced Crops: Biotechnology's Next Stage.* Washington, DC: Economic Research Service.

Lanterman, B. 2000. *Detection Methods for GM Plants and Plant Products.* Nepean, ON: Canadian Food Inspection Agency.

Organisation for Economic Co-operation and Development (OECD). 1993. *Safety Evaluation of Foods Produced by Modern Biotechnology: Concepts and Principles.* Paris: OECD.

_____ 1999. *Modern Biotechnology and the OECD.* Paris: OECD.

_____ 2000a. *GM Food Safety: Facts, Uncertainties and Assessment.* Report of OECD Edinburgh Conference on the Scientific Aspects of GM Foods. Paris: OECD.

_____ 2000b. *Overview and Compendium of International Organizations with Food Safety Activities.* Paris: OECD.

World Health Organization (WHO). 1991. *Strategies for Assessing the Safety of Foods Produced by Biotechnology.* Report of a Joint FAO/WHO Consultation. Geneva: World Health Organization.

# 5

# Industry Stewardship as a Response to Food Safety Concerns

*Lorne H. Hepworth*

Consumers, citizens and governments are demanding a much higher and more transparent level of safety in food production. The chemical industry is caught between consumers and governments, as they seek these new, higher standards. In some cases, governments or retail chains themselves can impose new requirements onto the food system, but more often producers and their suppliers and processors are the only actors with the capacity to develop or deliver new standards. This chapter reviews the contextual factors facing the food industry and examines two recent industry stewardship initiatives which are being undertaken to enhance risk management in the production and use of chemicals and biotechnological inputs.

The Crop Protection Institute (CPI) is the trade association that represents the manufacturers, developers, and distributors of plant-life science solutions for agriculture, forestry, and pest management — that is to say, pesticides and genetically modified (GM) crops. The time and costs associated with the research and development process, plus the regulatory review period, means it takes several years and substantial financial resources to bring a new product to the marketplace. Because of the nature of our products we are highly regulated by government. Our member companies include firms such as Aventis, Bayer, DowAgrosciences, Dupont, Monsanto, Novartis, and Pioneer HiBred. Virtually all of our members operate on a global basis.

Over the past few years our industry has redefined itself. Many of the companies have acquired equity positions in, or formed joint ventures, with seed, biotech, food, feed ingredient, and functional food companies. Thus, chemical companies that were once known for their innovative products have now redefined themselves by adding biotechnology as a means through which to develop new products. These go beyond our traditional crop protection markets — beyond weed killers, insecticides, and fungicides — to include other aspects of crop production, and novel food and feed production, which have elicited a variety of responses from the public.

## NEW REALITIES

Citizen and consumer perceptions are having an increasing influence on the market. During the 1990s, organic food production gained new respectability in Canada and elsewhere. Under the auspices of the Canadian General Standards Board, there now exists a standardized definition of what constitutes "organic" produce (similar efforts are underway in the United States and the European Union). Organic food production appeals to those who perceive that current farming practices are not good for them as individuals, the country or the planet. Some consumers are also concerned about the issue of "food miles." They draw attention to the energy, logistics, and social impact of transporting food over long distances, as opposed to growing and purchasing food locally. Various related "buy local" efforts fly in the face of the globalization agenda, or put another way, they present a trade barrier that is being erected by the individual consumer. Governments, world trade bodies, and corporations have found themselves powerless to deal with this new type of trade barrier.

What does the average member of the public make of all this? Based on Canada's excellent track record for supplying abundant, safe, nutritious, and wholesome foodstuffs, the issue of food safety should not even be on their radar screen. However, given the events of the last few years, it is not surprising that there do exist elements of uncertainty, confusion, and concern. Questions exist regarding the acceptance of food grown with new technologies like biotechnology, and with old technologies, such as pesticides.

As a result, a number of disconnects have developed. First, the firms developing new technologies are, on occasion, sending mixed messages themselves. For example, one division of a company may be selling GM seed, while the food division of the same company may declare that it will not use GM ingredients. Other well-known and respected companies have outright said "no" to genetically modified organisms (GMOs). Canadian-based McCain's, the world's largest producer of french fries, rejected Bt potatoes — potatoes that have built-in protection against the potato beetle — not on the basis that it represented a danger or a food safety threat, but because of the public's perception of the technology and its "bad PR." Although I may not agree with McCain's decision I respect their right to take such action. It must be recognized that when a company as highly regarded as McCain's takes such measures, it will not go unnoticed by the public who are likely to conclude that the technology must be flawed.

Second, regulators are sending diverging messages. When the public looks to the regulators charged with safeguarding their safety and the environment, they see conflicting messages in their approach to GMO regulation. For example, the European Union has taken one approach, a very restrictive one, but Japan has taken another approach and so has North America.

Third, society has multiple objectives, which can at times be complementary and at other times irreconcilable. Consequently, society itself is also responsible for projecting conflicting messages. To illustrate this point, it is worth returning to the example of McCain's. It has been suspected that recent fish deaths off Prince Edward Island are the result of the presence of a pesticide residue in soils from fields that had recently been sprayed. Torrential rainstorms then caused the pesticide residue to enter the river. The primary preventative measure has been to limit farmers from using fields near the edge of the river. Nevertheless, about the same time as the fish deaths, McCain's announced that they would not accept Bt potatoes. Shortly after, the CPI received a letter, jointly signed by PEI's ministers of agriculture and environment, asking the organization to encourage members to accelerate research on new chemistries that were less toxic to fish, and, where such products were available in the US, to get them registered for use in Canada. CPI replied that it is the constant goal of its members to research and commercialize lower risk technologies and, indeed, one of the newest of these was Bt potatoes. The advantage of Bt potatoes is that they give farmers another option, in addition to chemical insecticides, when managing the Colorado potato beetle. More options, more tools in the farmer's pest-management toolbox, are fundamental to Integrated Pest Management, which is a cornerstone of sustainable agricultural practices. If farmers utilized Bt technology, they could in some instances reduce the use of pesticides and minimize the risk of pesticide-contaminated soil and the risk to fish. Now, many in society have argued for farmers to reduce their use of pesticides and to examine alternatives for pest management. Unfortunately, some of those same people are just as vocal in their opposition to the use of GMO technology. I would argue that for those who oppose pesticides, there is a certain element of intellectual hypocrisy — or at a minimum, a double standard — in opposing Bt biotechnology applications. In the case of McCain's, farmers were precluded from using GMO technology. So what should have been a win-win-win scenario for the farmer, the consumer and the environment turned out to be a lose-lose-lose scenario.

Finally, consumers are not always consistent in their preferences. The public's apparent preparedness to accept "health" foods and herbal medicines without safety and efficacy tests is puzzling. This constitutes a direct contradiction of their expectations for greater assurances of safety for traditional and GMO foods. All the more fascinating is that their consumption of these often untested and largely unregulated products, pills, and concoctions is increasing markedly and attempts by government to put in place more stringent regulation has met with considerable opposition from some parts of the industry. One cannot help but think that the government is feeling pretty exposed in this area, and rightly so.

There is also an element of what some call "new activism" that has entered the anti-GMO food, anti-globalization, anti-multinational debate. We saw an example of this new dimension at the 1999 World Trade talks with the massive street demonstrations, since dubbed "The Battle of Seattle." Civil society engaged itself in a not-so-civil riot. The protestors were well organized, media savvy, and wired. They made wide use of the Internet and other information technologies to marshal their troops, keep the police off balance, and get their message to the public. For many watching the events on TV, it would be easy to dismiss this as mere hooliganism and not take it seriously. But to do so would be to underestimate what is happening; this new activism represents a potentially powerful force that neither industry nor government yet knows how to address.

Another reality is that the public's trust and confidence in science and science-based regulation has been tarnished — minimally so far in Canada, but nonetheless to some effect. In Europe it would appear that there has been a more significant erosion of trust and confidence in both science and the regulatory system in the aftermath of mad cow disease. Generally, people want an absence of risk in their lives, failing to realize that "zero risk" does not exist in any aspect of life. This desire for zero risk has manifested itself in the emergence of the "precautionary principle" or, as our industry prefers, the "precautionary approach." Anecdotal "possibilities," rather than scientifically-established "probabilities" and margins of safety, are challenging established approaches to risk assessment and risk management.

The consumer's quest for fresher, more wholesome, safer foodstuffs and, in some instances, grown in a way that reflects their social conscience will continue. Those desires, combined with crops being genetically modified to impart specific health or nutritional benefits (functional foods and nutraceuticals), or for vaccine production, will have a major impact on the entire food channel. The early applications of plant biotechnology have been of greatest value for farmers — traits such as herbicide tolerance and insect resistance — which the average consumer has had difficulty accepting. Certainly, crops engineered to grow in saline soils or drought-stricken areas that limited or precluded food crop production will be of great value, especially in sub-Saharan Africa. Nevertheless, these farmer-input traits are expected to represent only a small slice of the potential market value of all traits under development, a value that some estimates have placed at US$500 billion to US$1 trillion.

## APPROACHES TO A SAFE FOOD SYSTEM

There are two main areas that need more attention: the principles and guidelines that relate to improvements in the risk analysis process and a more active role for industry in voluntary initiatives in standards-setting.

Public policy should be based on science-backed, risk-assessment, and risk-management practices. While this is not a novel concept, having largely been the case in Canada and elsewhere, this view has recently been portrayed as outdated. Does this traditional approach ignore new socio-political realities that have entered into the equation? Does it deny what appears to be a greater desire for the public to be consulted and engaged in the process? Does this mean we cannot have a transparent and open system? No. Rather, the point is that society will not be well served if government policies and priorities are based on whimsical, anecdotal evidence, and "possibilities" of harm, rather than "probabilities" of what constitutes unacceptable risk. There is a real risk that priorities could be based on the lead news story of the day or on pet peeves that have a lot of political "sex appeal." Society will be well served if we use more scientific expertise in the examination of such issues. That is not to say that there is no role for multistakeholder involvement in questions of public policy, because there clearly is, but science must also be involved.

Part of the problem is that our current risk analysis model is focused on risk assessment and risk management. Too often, the third leg of this triangle, namely risk communication, or more properly, communication to the public of the risk-benefit equation, is overlooked. We need to de-mystify the science. We need to simplify our approach to explain not only the benefits of our technology, but also how the benefits exceed any of the risks incurred. We need to illustrate how our technologies benefit the public's quality of life. We need to demonstrate that the risks are minimal, or at least comparable, to the risks we have already accepted with existing production methods in many other parts of our life. There is no such thing as zero risk in any facet of life, and we need to put this into perspective for the general public. We need to do a better job of explaining how the regulatory process in place is open and transparent and is based on quantifiable science. We need to explain how this system protects the best interests of society in terms of measuring both the benefits and risks in the technologies put forward for registration. The North American Council for Biotechnology Information, for example, is a multinational initiative, funded by developers, to help provide the consumer with factual information on biotechnology, including its benefits.

In order to be effective, developers cannot be the only group communicating with the public; our regulators also need to communicate that we have a credible and proven risk-assessment and risk-management process that protects both our health and the health of the environment. We need our officials to say that we have a system based on science and measuring risk. We need our officials to say that when we utilize registered crop protection products to treat our food crops according to the guidelines that they have approved, there are no unacceptable risks for the benefits attained. To ensure the availability of our technology for our producers and stakeholders, and to ensure that we maintain an approval system based on science and not emotion, regulators need to share the responsibility of risk communications. Only then will the public have the reassurances they are asking for in terms of the safety and rigor of new products.

There is also a need to provide not only a response but also an alternative to the catchy ideas and critiques put forward by special interest groups. The critiques from advocacy groups like Greenpeace and the World Wildlife Fund have strong rhetorical and sometimes visual appeal when presented in the media or on the Internet. Often, however, their messages are merely vague public policy notions that lack substance. But in this new environment, these critics and their criticisms are often immune to the traditional science-based rebuttal our industry and government has relied upon.

In the past these groups were not as well organized and did not have access to the communications tools now available. Likewise, in the past, industry typically relied too heavily on the argument of science, using primarily reactive tactics, to preserve freedom to operate. Today, however, these groups must be taken very seriously as we debate both industry and policy objectives and strategy. They are able to organize and operate in any forum, using any tactic at any time and, often, are able to influence the public policy debate through their lobbying presence in the media. Industry needs to address the concerns of the general public and be more proactive in meeting environmental obligations through both our regulatory system and through industry stewardship initiatives. This requires acknowledging that there are changing public expectations and new realities governing how business is expected to be conducted. The issue of greater transparency in decision-making, and the capacity for public input into decision-making, even if it is not exercised, are important issues that need to be respected.

## INDUSTRY STEWARDSHIP

Just as government has a role to play in rigorous and robust legislative and regulatory oversight, so, too, does industry. Indeed, industry must show even more leadership in voluntarily accepting their responsibility to the environment and safety. Industry in Canada is working and investing in several targeted areas aimed at increasing the public's confidence and acceptance of new technologies. These include initiatives to foster "cradle-to-grave" stewardship of technologies, to support a high quality and efficient regulatory process, and to facilitate an improved and transparent communication process both with the public and between industry stakeholders.

The two main options are a new voluntary, self-regulating industry stewardship initiative and a "behind the farm gate" stewardship concept.

## Establishment of Industry Stewardship Programs

The Crop Protection Institute in Canada has taken a leadership role in terms of initiating and implementing, with member companies, two comprehensive stewardship programs: one for agri-chemicals and the other for biotechnology products. Both of these programs are under the *stewardshipFirst*™ umbrella. This program leads the world in its design and in its results to date. More than a continuously improving set of standards and codes, it is a guiding philosophy of proactive and evolving self-regulation. As an association, the CPI currently invests over 65 percent of its budget into *stewardshipFirst*™ programs.

The *stewardshipFirst*™ agri-chemical program encompasses initiatives that manage the stewardship of crop protection products through their full lifecycle in order to protect both the population and the environment. *stewardshipFirst*™ initiatives around crop protection products are illustrated in the following diagram.

FIGURE 1: Crop Protection Chemistry Stewardship

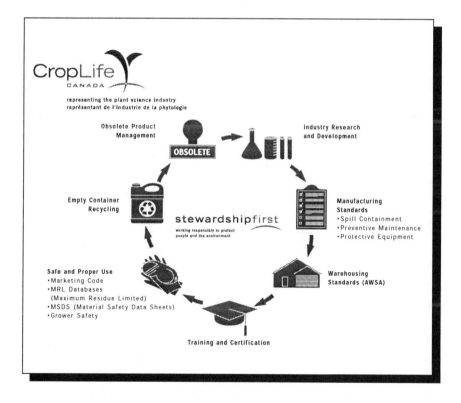

*Manufacturing Standards.* The crop protection industry's manufacturing standards assess and improve manufacturing processes in the areas of policies and procedures, safety and loss prevention, occupational health and environmental issues. Facilities formulating crop protection products must undergo an annual third-party, independent audit to ensure compliance with the established protocols.

*Warehousing Standards.* The warehousing standards program, managed by the Agri-chemical Warehousing Standards Association, is believed to be the largest self-regulatory initiative in Canada. The standards are designed to minimize the risks associated with the storage of agri-chemicals to employees, the public, and the environment. So far there have not been any major, or really, any minor incidents (e.g., fires, spills) involving the storage of products since the inception of this program. At the end of 2000 there were 1,791 warehouses certified through this program. With the support of member companies, the CPI has been able to enforce fully a "no certification, no ship" policy since the implementation of this program in March 1995.

*Marketing Code of Standards.* The marketing code represents the industry's commitment to self-regulation in the marketing of its products. It includes guidelines for ethical advertising and has a rigorous compliance and enforcement procedure.

*Certified Crop Protection Consultant.* This program, run in conjunction with the University of Guelph and the University of Saskatchewan, provides an opportunity for crop protection professionals to gain formal credit for their expertise and the opportunity to certify their skills in the stewardship of crop protection products. At the end of 2000 over 600 professionals in the industry were registered, and 388 have received certification.

*Grower Safety.* The *Pesticide Safety Handbook* contains guidelines for the safe use and handling of crop protection products. Featured topics include: the role of pesticides, good stewardship practices, managing risk, clothing and handling to protect, and storage and emergency response. Over 16,000 copies have been distributed to producers in all crop segments across the country. The CPI is also involved and strongly supports provincial grower certification programs across Canada. In many provinces, producers now require certification in order to both buy and apply crop protection products.

*Crop Protection Packaging.* Canada is recognized as a world leader with its empty pesticide container management program. Since 1989, over 34 million containers have been removed from the environment in Canada. In 1999 alone, the industry collected 4,419,000 or 62.2 percent of the containers shipped into the market from 1,031 collection sites. This is far and away the most successful program globally, and compares to a 42 percent collection rate in Germany and an estimated

26 percent in the United States. In Canada, the plastic containers are recycled into agricultural fence posts and highway guardrails, while metal is recycled into reinforcement bars for concrete.

*Obsolete Product Management.* This cross-Canada initiative provides a venue for the proper disposal of obsolete and unwanted agricultural pesticide products. This program is jointly funded by industry and the provincial governments, and is available at no cost to producers. As part of the *stewardshipFirst*™ lifecycle approach, this ensures that unused or unwanted products are properly disposed of rather than left to spoil the environment. To date, over 240 metric tonnes of unwanted product have been collected and disposed of safely. Logistically, this involves working with dealers and distributors at collection sites and having volunteers from industry staff and provincial environmental representatives on hand to process returned products.

*Maximum Residue Limit (MRL) Database.* As an extension of the stewardship activities, CPI is developing a database of Canadian pesticide MRLs and US tolerances (www.cropro.org). Agricultural trade and the export of crops like canola, wheat, and pulse crops, among others, are often affected by related trade barriers. This database will provide information that producers and channel partners need in order to make informed decisions about crop protection products for use on potential export commodities. This program, a key regulatory affairs initiative for the CPI, involves a comprehensive communications program targeting: Canadian producers (through producer organizations), crop protection retailers and distributors, member companies, processors and export companies, federal and provincial regulatory officials, and crop extension specialists. Beyond managing information on current MRLs, the broader industry goal is to support the harmonization of regulatory systems, especially between Canada and the United States to ensure a common product portfolio for producers on either side of the border. This would drastically reduce the need for other initiatives in this area.

The second stewardship program area is the Biostewardship initiative. This is a new area of endeavor for the CPI, and consequently, the scope and detail of this concept is still in the draft stage. As shown in Figure 2, it is intended to involve a very broad-based, multi-stakeholder approach using many of the same principles and approaches used in the successful implementation of the chemistry programs.

The first element to be implemented in the Biostewardship initiative is the development of a curriculum and training program for novel trait, confined field trial managers. The first training program will be for canola and consists of five modules. Pilot projects to test the curriculum were run in 2000, with full implementation scheduled for 2001. The objective is to reduce risk to the environment as it relates to field trials of GMO crops and help ensure compliance of field trial

FIGURE 2: Crop Protection Biotechnology Stewardship

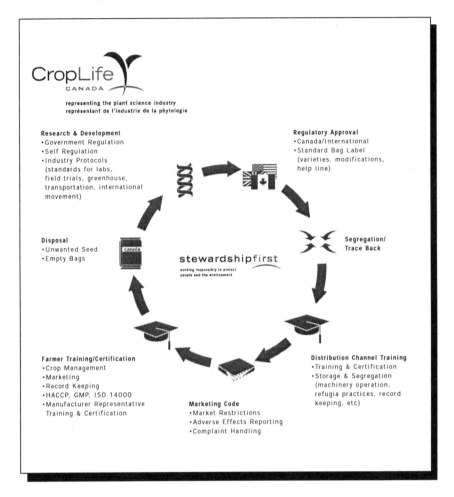

tests with government regulators. In succeeding years, other elements will be involved.

Industry has learned a lot through the initiation, research, development, and implementation of *stewardshipFirst*™ programs. First and foremost, as an industry association, we have had the trust, support, and most of all, the commitment of our member companies. We needed the commitment of not only their management teams in Canada, but internationally, to undertake programs like the ones we now manage. We enjoy a high level of unselfish commitment, dedication, and involvement at both the national and provincial levels within CPI. This has allowed us to utilize a vast pool of resources and expertise during the design and implementation of the programs.

Second, we have been very successful in working with industry stakeholder groups at all levels, from producer organizations through to our regulatory authorities. We have earned the trust of these stakeholder groups, and have returned that trust by listening and, with our member companies, investing in programs to protect our population and the environment. We have also learned that it is far more constructive to be proactive as an industry in developing these programs. Proactive, voluntary self-regulation has increased our credibility with not only regulatory officials, but also with the general public. This has helped to ensure our freedom to operate within our regulatory environment.

Third, tied to this is the importance of presenting a message that is consistent with other stakeholder groups involved in the production and distribution of our food supply. It is important to educate and provide timely and factual supporting information to these groups in order to have a coordinated and cohesive plan to address the issues before us.

## *"Behind-the-Farm-Gate" Stewardship Initiatives*

Finally, the agri-chemical industry is being proactive in recommending the adoption of "behind-the-farm-gate" stewardship initiatives. Canada's agri-food economy is built around the development of primarily international, but also domestic, value-added opportunities based on quality production. We have grown as an industry by providing our customers with the type and quality of product they are requesting.

To preserve and take advantage of our market opportunities, in both traditional markets and for new market opportunities, producers and their supporting infrastructures will need to adopt even more robust quality management systems to meet not only the rigours of increasing customer expectations and any regulatory requirements, but also to address the accountability needs of the general public. This may require programs to manage and track technologies from cradle to grave, or from seed to processing. At the farm-gate level, producers will begin to work toward Hazard Analysis Critical Control Points standards, Good Manufacturing Practices and International Organization for Standardization designations to qualify for identity-preserved contracts with income-generating opportunities. Producers in many market niches, whether in food-quality, export soybeans, canola, lentils, potatoes or sugar beets, have already adopted sanitation and record-keeping practices to meet customer and processor demands. Quality designations will be a help in overcoming commodity price malaise at the producer level and will be a major area of management investment over the next decade. The crop industry will be among the major stakeholders driving this process. Let there be no mistake that when a company comes looking for farmers to grow a crop — not for its protein or fibre value, but for the value of the vaccine that is going to be extracted — there is no doubt that they will expect the same kind of quality assurance "behind the farm gate" as they expect today in their drug manufacturing plants.

This will involve a restructuring of primary agriculture. Such a reorientation will likely include the use of Global Positioning System/Global Information System technology applications to provide for the more precise application of pesticides and fertilizers benefiting both the environment and the farmers' bottom line, and to document quality assurance measures. Segregation, trace back, and labelling systems will all be part of this process. That is why the debate over labelling of GMO foods may be somewhat moot. Industry has gotten itself on the wrong side of this one. It is perceived to be against labelling, ergo against consumer choice, when in fact industry supports consumer choice just as it supports farmer choice. The issue of labelling is moot because once there are output traits commercialized, the companies are going to want to label them, otherwise the consumer will not know to purchase them. A result of this will be greater segregation and identity preservation, well beyond what is going on already. This will require the development of new business arrangements in production agriculture, an area that farmers could use some help in developing.

"Behind-the-farm-gate" stewardship practices will be a tremendous undertaking. It will require a coordinated effort involving all sectors of the food industry. If successful, these practices present the opportunity for the food industry to address the concerns and demands of consumers with respect to food safety and quality assurance.

## CONCLUSION

Industry has recognized the legitimate concerns and demands of consumers, citizens, and governments for safe and nutritious foods and has responded in a number of proactive and progressive ways. The private sector can and should play a role in meeting and accepting its responsibility for public safety and to the environment — as we say, putting stewardship *first*.

In Canada, the crop protection industry believes that the technologies we develop are critical to the survival of our civilization both now and into the future. We know in industry that we have a good story to tell, and need to do a better job of telling it within an ever-changing environment of social acceptance. We also understand that being advocates for, and in many cases defending, our technologies will not become an easier task anytime soon. But we realize that our actions and our efforts in the stewardship and communication areas are a critical aspect of risk management and are vital to the acceptance and continued availability of our products.

# 6

# Risk, Precaution and the Food Business

*Neville Craddock*

## INTRODUCTION

Most of the food we eat is processed to some degree after it leaves the farm and before it arrives in our kitchens. Examples range from simple, washed, and chopped salads to highly sophisticated ready-to-eat meals, reflecting the widest spectrum of ethnic recipes from around the world — available to the consumer in frozen, chilled or ambient-stable formats. In the United Kingdom alone, it is estimated that there are some 600,000 food businesses, ranging from the small corner butcher who makes his own sausages to giant firms with operations in many countries.

My employer, Nestlé S.A., based in Switzerland, is the world's largest food company, functioning in virtually every country across the world, employing some 225,000 people, operating over 500 factories, and producing about 15,000 different products in every sector of the food and beverage industry. It also has substantial interests in pet foods, cosmetics, clinical nutrition, and health-care products. Our business depends on being responsive not only to the public's concerns about safety, and indeed about all aspects of their food such as its origins and composition, but also to the regulatory frameworks created by governments around the world. The company, therefore, has an active interest in all aspects of current food science, safety, and international trade. My personal responsibilities cover regulatory and environmental affairs, as well as the scientific and technical issues surrounding them, within the UK business of Nestlé S.A. I am a scientist by training, but I have spent the last 20 years increasingly dealing with food legislation.

In this chapter I discuss the interaction between the food scares we have experienced and the regulatory mechanisms necessary to prevent future concerns. I divide the discussion into four main areas: (i) the implications of the globalization of food and of recent food scares; (ii) modern risk analysis and commerce; (iii) a brief review of the precautionary principle and how it might realistically be applied; and (iv) possible ways forward to address the current concerns.

## GLOBALIZATION OF FOOD AND FOOD SCARES

The food industry and its products are incredibly diverse, with traditional production and manufacturing methods existing alongside an ever-increasing range of new technologies and processes that are producing an ever-widening range of food products. All the major food companies source their raw materials and, increasingly a significant proportion of finished products, on an international basis. Products made in any given factory are as likely to be destined for export as for consumption on the domestic market. Equally, products purchased by consumers are very likely to have been produced wholly, or at least in part, in an overseas factory.

Increasingly, therefore, we need to perceive the widest possible horizons as a single trading entity. This requirement for all parties — industry, regulators, and consumers — to think globally has two main policy implications. First, our industry and its ability to deliver safe, consistent, quality products to consumers depends more than ever on stable and predictable trade regimes for both raw materials and finished products. The World Trade Organization (WTO) negotiations on agricultural trade are vital; further liberalization will reduce the burden of protectionist farm subsidies on consumers and taxpayers. It will also help to smooth shocks in supply and demand thanks to a freer, less distorted international flow of food. Second, in the area of food safety, it is becoming increasingly apparent that the need for a single, equivalent framework of good hygiene practices and equitable, enforceable rules is paramount. Increased international trade must be accompanied by schemes able to respond rationally to consumer requests (for example, the assurance of food safety throughout the whole food chain), but also to avoid a public backlash against market opening and consumer concerns being used as a pretext for protectionism. The right framework is being constructed with the WTO Agreement on the Application of Sanitary and Phytosanitary Measures (SPS) and its close relations with the work of the *Codex Alimentarius* Commission (see chapters by Anne Mackenzie and Peter Phillips).

From a purely commercial point of view, we know that people with a per capita income around US$6,000 spend a higher-than-average share of their disposable income on manufactured food. Higher growth in countries close to this average income will therefore have a significant impact on food sales and, potentially, on international trade in food.

In recent years, numerous food scares in the UK and Europe have reduced consumer trust in science, the public authorities (who have reacted with strong, but not always well-conceived actions), and even the regulatory mechanisms themselves. Issues such as bovine spongiform encephalopathy (BSE), E. coli, listeria, salmonella, dioxins, and various other food contaminants have been fuelled by emotional overtones, such as challenges to the "ethics" and "naturalness" of modern agricultural and food production, that have not contributed to an understanding of the real issues at stake. Although many of these food safety incidents and concerns have been derived from agricultural raw materials and packaging — areas not under the direct control of the manufacturing industry — the end result, nevertheless, has been that it is the food industry itself that is perceived as less than perfect in the minds of the consumer.

The BSE crisis, initially in the UK but more recently in several other countries, added a new dimension to the discussions, leading to a fundamental questioning of the competence and independence of public authorities, the effectiveness of the control system in protecting the consumer, and in the ability of the food industry itself to deliver safe food. The inability of science to provide definitive answers to all the questions surrounding food safety incidents has not helped. Many of the challenges have been, in effect, to prove absolute safety — or put the other way round, to prove the negative, that is, that no harm or damage will ensue from a particular decision or action. This, of course, is fundamentally impossible.

More recently, the concerns have been extended to encompass even broader and less specific aspects of food production. The debate on the growth and use of genetically modified crops and their highly purified derivatives in food production covers not only food safety but also environmental safety, the ethical aspects of "playing God" with the development of this technology and even the political issues of the increasing global domination of world agriculture by a few multinational organizations. However, as in the early days of BSE, many of the questions asked do not have validated scientific answers.

We are seeing these "safety concerns" rising around the world — fuelled not only by the media and in many cases politically motivated lobby groups, but also greatly facilitated by the rapid transfer of information via new electronic systems. Yet, the truth is that food has never been safer: many manufacturing premises require licences to operate and are subject to routine on-site enforcement; sophisticated quality systems are in place; modern distribution is fast and efficient and frequently under strict temperature control; and supermarkets offer high standards of hygiene and cold storage.

Taken together, these pressures have resulted in a changed political framework whereby, in Europe at least, the consumer is now "king," and is strongly supported in their responses to various food scares by the major retailers who control some 60 percent of food sales. This has resulted in a situation in which multinational manufacturers have sometimes been obliged to adopt local policies in relation to issues (such as genetic modification) that may not be wholly compatible with their longer term global views.

Nevertheless, despite all the demands and pressures, we must not forget that 100 percent safety or zero risk simply does not exist in the real world. Furthermore, all control systems are ultimately designed and operated by human beings, and despite their best endeavours, mistakes and accidents will happen. It is therefore incumbent upon industry and public authorities to ensure that appropriate systems are in place to minimize any risk to consumers in the event of such accidents, and to persuade the public at large that the situation remains under control.

In response to the demands for greater consumer involvement in all aspects of food control, "openness" and "transparency" have become key words. A further response is that the "precautionary principle" is being quoted or invoked much more frequently and developing, applying, and explaining the concept of risk analysis — the assessment, management, and communication of risk — appears almost to have become a university degree in itself.

## A COMMERCIAL PERSPECTIVE ON RISK ANALYSIS

There is a growing demand that food should always be perfectly safe to eat — there should be zero risk. This is unrealistic in the real world and is not a philosophy that is applied in non-food areas. Zero risk does not exist. While it will be difficult to convince the consumer of this, it has to be done. New approaches to risk management, in areas where scientific uncertainty exists and there is no possibility of carrying out a comprehensive risk assessment, will be critical and must reflect the need to communicate these realities to consumers.

Considerable work is therefore being done on: how, scientifically, to assess the degree of risk in any particular area; how to manage the risk (including the political, as well as the scientific considerations); and, perhaps most importantly, how to communicate the facts about the risk involved to consumers and to inform them that it is impossible to eliminate risk completely.

Consumers' overall perceptions are essential to all aspects of product acceptability and for the effective functioning of markets. The difficulty is to ensure that these perceptions reflect a balance of all available scientific evidence and do not, through misrepresentation, lead to an undermining of the essential science-based approach.

Responses to change can easily be driven by misperceptions that obscure reality. Individual consumer attitudes toward new processes, new technologies and new products may be strongly influenced by the media or by public scepticism of scientific evidence and the motives of those involved.

However, while we are all fully aware of the need to react to consumer pressure, an inappropriate response to a disproportionate perception of risk may be counterproductive, deflecting attention from more important areas of higher, but less emotive and less publicized, concern. This can precipitate changes to policy and organization that focus only on the short-term problem, but fail to address and at times even undermine the environment necessary to develop longer term solutions to specific issues.

We are therefore seeing a dramatic increase in the UK and Europe of transparency and independence in the mechanisms whereby scientific decisions, and the necessary regulatory consequences, are established. In this way, it is hoped that the public's confidence in the system of legislation and control will be restored, and that consequently, their confidence in the ability of the food industry to deliver foods that are as safe as realistically possible will return. In Europe, various committees of experts, appointed for their excellence and profile in a particular field, already provide the scientific advice to underpin the legislative process. The independence and robustness of their opinions constitute key contributing factors to the credibility of legislation. The key to future consumer confidence will be to ensure that this independence and robustness is not only recognized but maintained in the face of mounting criticism.

There are also many international examples of the involvement of all interested parties in the formulation of legislation. These procedures have been accompanied by requirements for extensive public consultation before enactment, often reinforced with an obligation to satisfactorily address every special concern. The experiences of the United States, Australia, New Zealand, and other countries offer impressive examples.

Nevertheless, the principle of decisions based on sound science, may be a less straightforward requirement than it appears. For scientific advice to the legislator to be useful, it has to meet three essential criteria: it has to be relevant, reliable, and rapid.

*Relevance* requires that questions are appropriately formulated, so that the answers obtained are sufficiently precise and focused. This may not always be easy. For example, asking whether a particular food, ingredient or process is fully safe, meaning zero risk, would almost invariably produce a negative, or at least a highly qualified, answer. On the other hand, an alternative question, based on a definition of acceptable levels of risk, could be seen as pre-judging the issue.

*Reliability* may not be easy to achieve. Available evidence may not be sufficient to evaluate the possible risks, particularly in the case of new technologies or previously unknown issues.

It may not be easy to obtain *rapid* advice that is sufficiently timely for the decision-making process. In some cases, legislation requires specific decisions to be taken before particular products can be marketed, but does not specify a fixed time limit. This can lead to commercially serious delays.

Serious problems will arise, in particular, at the international level when local legislation is out of line with that in other countries. Inconsistencies, unnecessary detail and rigid procedures may well reflect the difficulties of reconciling internationally divergent views on issues of serious public concern. But they do not facilitate compliance; they increase costs and can be a source of trade barriers. In some cases, decision-taking has become very slow and cumbersome; genetically modified organisms (GMOs) are a particularly pertinent example. However, although these delays appear to be an unnecessary barrier to trade, they may in reality reflect a difficulty in finding acceptable local compromises.

Equally, legislators need access to the widest possible range of scientific expertise in order to obtain the best and most up-to-date advice. This will be particularly important if local measures that go beyond those of the *Codex Alimentarius* Commission are proposed and if these are to withstand international scrutiny.

However, science is not a static discipline. Opinions and advice always contain an element of uncertainty. Hazards may emerge where they did not previously exist. Evolving analytical techniques and refinements to microbiological knowledge mean that scientific advice, and measures based on it, need to be reviewed from time to time.

Sometimes, when new hazards have been identified, there may be little reliable scientific data available. Nevertheless, this does not exempt industry, authorities or governments from their responsibility to introduce measures to protect consumers. If severe risks are theoretically possible, it may be appropriate to apply the precautionary principle; however, this should be confined to clearly defined circumstances.

The key question, of course, is how to determine what is an acceptable level of risk in any given circumstance, when all other legitimate factors have been taken into account. This is the decision of politicians. On the one hand, the wrong decision can have a serious, adverse impact on the economy, reducing the potential for industry to use or develop new technologies or to introduce foods accepted in other parts of the world, and ultimately reducing consumer choice. On the other hand, the wrong decision may fail to accommodate the legitimate concerns of consumers or those with a less economically driven approach.

Current food legislation reflects a blend of scientific, societal, political, and economic forces. Whilst many provisions were developed in a coordinated manner, others have been introduced in an ad hoc way, in response to the particular issue of the day. This has resulted in a marked lack of consistency, with some legislation prescribing every last detail, without paying enough attention to the principle of proportionality, while other legislation is drafted in such general terms that its interpretation becomes highly subjective and its enforcement inconsistent.

There is clearly considerable scope for global simplification of food legislation; however, industry would be the first to acknowledge that this must not result in any reduction in levels of consumer protection or product safety. General safety requirements for all foods should not be differentiated. The essential prerequisite is not that the products themselves are defined, but that their manufacturing and handling processes should be the key factor in determining the way in which legislation is applied to them.

Modern food safety and hygiene legislation is increasingly being based on the principles of Hazard Analysis Critical Control Points (HACCP). This is a broadly applicable quality management tool, developed over the past 30 years by the NASA space program that entails a thorough evaluation of the risks associated with every step in the production chain. Every step that may present a risk is identified and particular attention paid to those that are considered to be "critical" to the safety of the final product. A proper HACCP-based analysis will take into account all chemical and microbiological hazards for the final consumer, including those that may occur after manufacturing and marketing. Implementing HACCP is primarily the task of the responsible manufacturer and distributor, with appropriate input from the competent authority.

However, certain fundamental, practical issues must be recognized. First, although it becomes relatively easy to specify in legislation the objectives to be met, and to place the responsibility on industry to define the means of achieving them, many (if not most) smaller businesses — and, indeed, many enforcement officers — will not have the appropriate expertise or resources to carry out full HACCP analyses. Second, any system for the protection of public safety that is based on subjectively worded, general legislative requirements is dependent on a uniform and consistent legal interpretation, implementation, and enforcement for its success. Finally, one major objective behind the rethinking of food safety control mechanisms is to restore consumer confidence. But quality assured by strict adherence to the process rather than by a government inspector often bothers consumers. Regrettably, consumer confidence in industry's ability to police itself is not high at the present time and it may be several years before a HACCP-based system becomes globally acceptable.

The skill of the legislators will be to provide an acceptable balance between meeting the consumer-led demands with the requirement to meet international trading obligations. These difficulties should not be underestimated.

## RISK AND PRECAUTION

Reference was made earlier to the fundamental principle that legislation should be based on sound science. However, in the current political climate, it is my personal opinion that this long-established principle today carries a slightly different emphasis from previously. The legislation is likely in practice only to be *based on*, but not *driven by*, sound science and there is a greater questioning as to precisely what constitutes *sound* science. In addition, even when science is recognized as the cornerstone, there may be perfectly valid reasons, based on ethical, religious, or environmental considerations, or arising from specific production methods, to apply more protective measures than purely scientific evidence suggests is necessary.

Following the BSE crisis in particular — where, first, the disease was not originally known to veterinarians and, second, for nearly ten years the scientific experts stated clearly that it could not cross the species barrier into humans — we must recognize that it is necessary today to strike a balance between scientifically identifiable risks and what society will accept. Other incidents outside the food industry that have similarly reduced the credibility of scientists and helped to fuel consumer criticism of control mechanisms include the accidents at Three Mile Island and Chernobyl, or Bhopal and Seveso. Thalidomide would be a further example of a failure of science to predict disastrous consequences. This management of risk urgently requires the development of internationally accepted methods for determining acceptable risk and safety objectives around which legislative targets can be set. Hence the ongoing global discussions.

So what exactly does the precautionary principle imply?

Recent events have highlighted an increased sensitivity to potential risks allegedly associated with modern foods and their production methods. For example, wide-ranging concerns have been expressed about animal welfare in intensive rearing systems, the spread of antibiotic resistance, the spread of various animal and human diseases and even questioning the "naturalness" of the foods produced by intensive agricultural systems. In some cases, this sensitivity has been apparent, and frequently fuelled by the mass media, before scientific research has been able to explain fully the problem in question. Public opinion then pushes political decisionmakers to recognize these fears and to adopt potentially disproportionate, preventive measures to eliminate the risk, or reduce it, and the fears, to an acceptable level. As stated previously, the wrong decision can have a serious, adverse commercial impact and/or may fail to accommodate the legitimate concerns of consumers and others.

In practice, the "use of the precautionary principle" is being claimed as the justification of the choices made by decisionmakers responsible for the safety and welfare of fellow citizens — political decisions that are made in conditions of scientific uncertainty. Between the extremes of banning something until science has proved its complete harmlessness (a scientific impossibility), and not banning something until a real risk has been proven (which is likely to be politically unacceptable), there is a wide spectrum within which to apply or mis-apply the precautionary principle.

Globally, the link between the necessary degree of certainty in the science behind the risk assessment, and the application of the precautionary principle has yet to be fully defined. There is therefore a need to establish clear working guidelines if its application is not to become an increasing area of contention in the future.

The precautionary principle has been well developed in the environmental field following the Rio Convention and has, in general terms, been enshrined in the SPS Agreement. The EU Commission stipulated in its 1997 Communication on Consumer Health and Food Safety, that it would be guided in its risk analysis by the precautionary principle where the scientific basis is insufficient or uncertainty exists.

Article 5.2 of the SPS Agreement makes it clear that non-quantifiable data of a factual or qualitative nature may form part of a risk assessment — hence the ongoing debate between the EU and the United States on growth-promoting hormones. The EU has taken a slightly different working definition of the precautionary principle, defining it as: "an approach to risk management applied in circumstances of scientific uncertainty, reflecting the need to take action in the face of a potentially serious risk without awaiting the results of scientific research." Notably absent from this definition is a specific reference to any time period in which the necessary, relevant information should be sought and by whom.

So, when and how should risk managers decide to take a precautionary approach? It is possible to set down a clear framework of principles that should be followed:

- Start with an objective risk assessment, identifying at each stage the degree of scientific uncertainty;

- Involve all stakeholders to study the various management options that may be envisaged once results of the risk assessment are available and make the procedure as transparent as possible;

- The measures proposed must be proportionate to the risk that is to be limited or eliminated;

- The measures must include, at an early stage, a cost-benefit assessment (i.e., an assessment of the relative advantages and disadvantages) with an eye to reducing the risk to a level acceptable to all stakeholders;

- The authorities must establish a clear responsibility as to who must furnish the scientific proof needed for a full risk assessment as well as the appropriate time-scale; and

- The measures must always be of a provisional nature, pending the results of any scientific research to furnish the missing data and the performance of a more objective risk assessment.

## THE WAY FORWARD

The international legislative process is extremely complex and it can take several years to pass from concept to implementation. Of course, safeguard measures exist to cover immediate dangers to health and these can be adopted rapidly if necessary. However, there is an urgent need and considerable scope to streamline the regulatory systems, and to improve both the availability and timeliness of scientific advice.

Practices in some individual countries point to effective approaches that could be considered for a wider application. For example, in the United States, the Food and Drug Administration (FDA), operating through its Centre for Food Safety and Applied Nutrition, develops and applies food legislation, from horizontal measures to detailed approvals, including pre-market approvals, if appropriate. Acting as an agency, it is advised by its own scientists, as well as by other advisory committees. Approval is administrative and does not require Congress to regulate. Australia and Canada have similar mechanisms.

Significantly, in all these cases, there seems to be a higher level of public confidence in the system for food legislation and control than is currently the case in the UK or Europe. The main strength appears to derive from their political independence in that, while all interest groups are kept fully informed and consulted, they do not directly influence specific decisions — or at least are not seen to!

While the problems and critical aspects of the present system of food legislation are well identified, no easy, globally applicable solution is available. However, in any discussion of possible reforms, it is essential to keep a sense of proportion, recognizing that current systems have generally been successful in ensuring a high level of safety and consumer protection and, equally, avoiding international trade barriers. Nevertheless, recent events have exposed weaknesses in the system that need to be addressed if we are to strengthen consumer confidence and to dispel emerging threats to international trade.

A number of basic goals and fundamental principles for food legislation can be identified as it must provide a high level of protection of consumer health and safety whilst being proportionate and effective. It must encourage innovation and market entry, and the free movement of goods, thus supporting a competitive industry, with strong export potential. It must support and encourage rather than prevent the supply of safe, wholesome foods. Equally, it must not impede responses to changing consumer demands and it must be capable of being adapted easily to scientific and other technological developments.

Legislative controls should primarily be based on sound scientific evidence and risk assessment, but must be sufficiently flexible to reflect ongoing practicalities and be proportionate to their aims. They should place the primary responsibility for safe food on industry, using HACCP-type systems backed up by effective and efficient official control and enforcement.

Legislation must be written in a coherent, rational, and user-friendly form that is capable of uniform interpretation and equitable enforcement across its range of application. Providing this is done, there should be no undue impediment to the free movement of foodstuffs or imbalance of impact on any sector of the industry.

A key element for consumer confidence is that, where legislation is in place, it should be fully and uniformly enforced. Consistent and equitable interpretation and full implementation and enforcement are essential if all parties are to have full confidence in internationally-traded products.

The complexity of food production and distribution argues for an even-handed and proportionate approach that covers the whole food chain. The strength of a chain depends on all the links, and there is therefore a clear need for integration of measures from farm to consumer. There is no point in restrictively controlling one part of the food chain if control is more effectively or, conversely, inadequately exercised elsewhere.

Food safety legislation should therefore apply throughout the chain, from farm to fork, to use the current European jargon. It should impose the responsibility for the safety of food and the prevention of unacceptable risks to the consumer onto industry, with requirements based on goals, intended results, good hygiene practices, and HACCP principles. It should grant industry flexibility to meet its obligations through the most appropriate means available, and thus to respond quickly to new hazards while providing a basis for innovation.

It is frequently claimed that HACCP is "only for the big guys" and is too complex and bureaucratic for the small businesses to apply. I refute this totally. While there is undoubtedly a real danger of a secondary, excessively bureaucratic auditing and control industry being created in the present climate, I would maintain that much of this is unnecessary. In reality, HACCP should be considered a responsible and proportionate use of common sense in the day-to-day management of the risks associated with a given process or operation. The ability to demonstrate an adequate knowledge and understanding of the basic food safety risks should be a fundamental requirement for any business, large or small, wishing to operate in the food industry. I would go further and question whether any food business operator who could not demonstrate such an understanding should be permitted to continue to operate in our industry. There should not, however, be an automatic, statutory requirement for low-risk businesses to be forced into excessive and ongoing bureaucratic record-keeping.

This challenges industry, particularly individual managers within smaller businesses, to maintain a good technical understanding of food safety. It also reflects recent international trends whereby industry takes the responsibility for the control of the foodstuffs it produces, backed up by official control systems. However, in the light of recent experiences with food safety issues, it is clear that many interested parties are not yet ready to accept HACCP as the legal mechanism for food safety control.

The consumer is entitled to expect, and to receive, equivalent guarantees of safe food wherever it is made and wherever it is purchased. Food safety legislation must therefore be applicable to all relevant stages of the food chain, regardless of the size of the enterprise. Derogations from food safety controls should be minimal, if any, and, if granted, must in no way prejudice consumer confidence in the totality of the control over safety aspects, whether exercised by control authorities or the industry itself. Where there are concerns that the burden of legislation may be excessive for any scale or location of an individual food business, the appropriate application of HACCP will, in fact, serve to identify precisely which elements of the legislation are relevant to ensuring consumer safety and those that are not. In this way, it will then become possible to focus attention on the critical elements of the business in a manner that is directly proportionate to the food safety risk. The application of HACCP, rather than being an overbearing imposition on the small business, will thus *reduce* the impact of any legislation to the minimum necessary to ensure consumer safety.

In looking to review, streamline, and rationalize food safety legislation, it is possible to identify improvements in a number of areas. The legislation needs to be reviewed in the context of framework concepts, to define general measures that will ensure that all operators have a statutory duty to market safe and wholesome food. The focus must be on the need for, and proportionality of, specific legislation. There are strong arguments for a greater reliance on horizontal measures, setting out the aims to be achieved, rather than the precise methods to be used. This could remove many detailed provisions that restrict innovation and provide industry with the scope to choose between alternative ways to reach the same objective, while not in any way detracting from food safety. In cases where the technology used may be of potential ethical or other concern to consumers — such as irradiation and genetic modification — it may be necessary to provide appropriate information (e.g., through labelling) to enable them to make an informed choice about their purchases.

Following a farm-to-fork approach, food safety measures must be consistent and proportionate throughout the production chain. Unless it is absolutely unavoidable, there should be only one specific regulatory control on each particular food safety matter, whether for the approval of new substances for food use or to control naturally occurring hazards, that is, the "one door, one key" approach. The mechanisms for providing and assessing scientific advice need to address the three Rs: relevance, reliability, and rapidity. We need to address such questions as whether scientific advice could be more efficiently obtained, making better use of the many centres of scientific excellence around the world; and how scientific expertise and the legislative process can be better integrated, without compromising their independence.

The mechanisms for implementation of new legislation and for introducing technical modifications that are not fundamental to the underlying nature of existing legislation must be accelerated. Furthermore, consumers must be more fully integrated into the process, thus improving their acceptance of decisions.

There is no doubt that reactions to recent chemical or microbiological incidents, or more widespread issues such as BSE and the debate on genetically modified crops, demonstrate that the average European consumer has little knowledge about the food supply chain and little confidence in food safety procedures. In the UK and elsewhere, the average consumer is becoming increasingly divorced from rural life and, consequently, from the realities and practicalities of modern agricultural methods and food production. In addition, there is limited awareness of which (if any) institutions have the main and final responsibility for food safety standards, new product authorizations, and supervision of national compliance systems. This is at the root of the widespread loss of consumer confidence. It is really little wonder that the public appears to be so receptive of each and every potential food scare. Independent of any changes to systems of regulatory control, there is, therefore, an urgent need for the development and introduction of comprehensive information and education programs to address all aspects of the food supply chain and food safety control, not least as part of the formal school curriculum.

## CONCLUSION

Much has been said and written about the need for legislation to be science-based. It is essential that future regulation must remain based on scientific considerations, but industry would be deluding itself if it ignored the genuine consumer concerns that result in science-based conclusions being tempered by a political recognition of society's sensitivities toward certain issues. Equally, it is important to recognize that the precautionary principle will increasingly be applied, although it must, conversely, be recognized that nothing in this world is without some degree of risk. Nevertheless, it is essential for the profitability of the food industry, and consequently to the overall benefit of all citizens — whether carrying out fundamental research, investing in or employed by individual companies, or even pensioners — that industry should be free to compete on the global stage equally with all competitors.

# Safety: Risk Communication

# 7

# Consumer Perceptions of Food Safety: Survey Research in Economics and Social Psychology

*Spencer Henson*

## INTRODUCTION

The aim of this chapter is to explore the level and nature of consumer concerns about food safety in Europe, drawing predominantly on economics-based research, or at least research that starts out with the perspective of an economist.[1] In so doing, however, it is acknowledged that other social science disciplines, for example, social psychology, have much (and some would say more) to contribute to our understanding of consumer concerns about food safety (see, e.g., Bredahl, Grunert and Frewer 1998; Fife-Schaw and Rowe 1996; Frewer *et al.* 1997; Sparks and Shepherd 1994). Key issues that are addressed include the level and nature of consumer concerns about food safety, the ability of consumers to judge safety at the point of purchase, and the impact on food choice behaviour. Throughout, problems associated with consumer research on food safety are highlighted.

For many researchers Europe is an interesting case study. In its recent history, Europe has been subject to a number of "food scares" that have received wide-scale media attention. These have served to highlight the nature of the modern food supply chain and the practices it employs, of which many consumers were unaware or chose to ignore. Furthermore, they have demonstrated that modern science and technology, coupled with government regulation, cannot guarantee that the risks associated with food are kept within acceptable bounds. Given this recent history, it is not unrealistic to suggest that consumer concerns about food safety are more acute in Europe than in other developed countries (e.g., United States and Canada). Indeed, this assertion is supported by trends in food consumption, growth in demand for organic food products, and political pressure for reform of the public institutions that regulate food safety. Nevertheless, there is a paucity of research that actually compares the level and nature of concerns about food safety in Europe and other parts of the world in a rigorous manner. Consequently, much of the following discussion considers the nature of food safety concerns within Europe, or at least reports research that has been undertaken in Europe, rather than attempting a comparison with other countries.

## FOOD SAFETY CONCERNS IN THE UNITED KINGDOM

Numerous studies have explored the nature and/or level of consumer concerns about food safety in Europe in general, and the UK in particular, both generally and with reference to specific food products (see, e.g., Becker 2000; Frewer *et al.* 1996; Glitsch 2000; Henson and Northen 2000; INRA 1998; Latouche, Rainelli and Vermersch 1998; Saba, Rosati and Vassallo 2000; Taylor Nelson 2001; von Alvensleben 1997). Likewise, studies have been undertaken in other developed countries such as the United States (see Adu-Nyako and Thompson 1999; Baker 1999; Eom 1994; FMI 1999, 2000; Frenzen *et al.* 2000; McGuirk, Preston and McCormick 1990; Wang, Halbrendt and Caron 1996). The results of these studies are difficult to compare across countries and over time due to inconsistencies in data collection methods. Collectively, however, they indicate that consumers (or at least certain consumers) within developed countries in general have concerns about a variety of food safety issues, relating both to the potential hazards associated with food itself and the methods and technologies employed in food production.

By way of example, Figure 1 reports the results of a random postal survey (n=1,000) of food purchasers in the UK and US undertaken in 1999 (Henson forthcoming).[2] Respondents were presented with a range of food issues and asked to indicate those that were of concern to them when buying food for themselves and/or their households. The results of this study are of particular interest because they permit some form of European/non-European comparison.

While the foremost concern in both the UK and US was food poisoning (of concern to at least 70 percent of respondents in both countries), there were significant differences in the importance of other food safety issues. In the UK, prominent concerns included bovine spongiform encephalopathy (BSE) and use of pesticides in food production, whereas in the United States the fat content of food and use of additives in food were of greatest concern. Overall, there was a small but significant difference between UK and US respondents in the level of concern across these food safety issues — the average level of concern was 54.9 percent in the UK and 51.2 percent in the US. Likewise, there was a small but significant

FIGURE 1: Prompted Concern about Food-Related Issues in UK and US, 1999

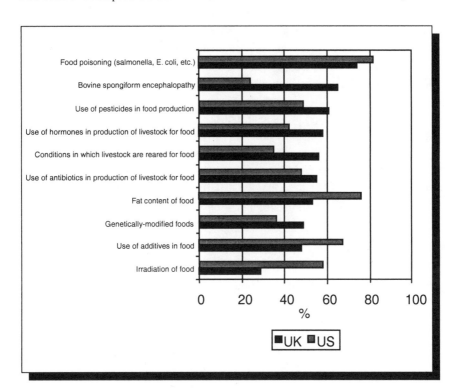

Note: Level of concern in UK and US statistically significant at 5 percent level except for food poisoning and use of antibiotics in production of livestock for food.

Source: Henson (forthcoming).

difference in the proportion of respondents that were not concerned about any of the issues presented to them: 7.3 percent in the UK compared to 10.7 percent in the US.

The results reported in Figure 1 suggest that, whilst the overall level of concern about food safety is quite similar in the UK and the US, there are differences in the issues that are to the forefront of the consumer's mind. However, while these results may provide some broad indication of consumer concerns, they are very much a facet of the method of data collection. More specifically, prompted questions of this type tend to provide an exaggerated measure of the degree to which consumers are actually concerned about food safety issues.[3] By way of comparison, for example, Figure 2 presents results from the same study, where consumers were asked to identify, unprompted, those issues associated with food that were of concern to them when buying food for themselves and/or their households.[4] The results are quite different, both in terms of the level of measured concern, which is consistently lower than the prompted responses, and the ranking of individual issues. For example, while 29 percent of respondents claimed to be concerned about irradiation of food when promoted, only 3 percent mentioned this issue unprompted.

A wide range of economic, social, political, and cultural factors may help to explain differences in consumer perceptions of the risks associated with food and their responses to new risk information. A full review of these factors is beyond the scope of this chapter, although it is useful to consider at least one or two of the underlying reasons why there may be differences in the level and/or nature of consumer concerns between Europe and other developed countries. On the one

FIGURE 2: Unprompted Concern about Food-Related Issues in the UK, 1999

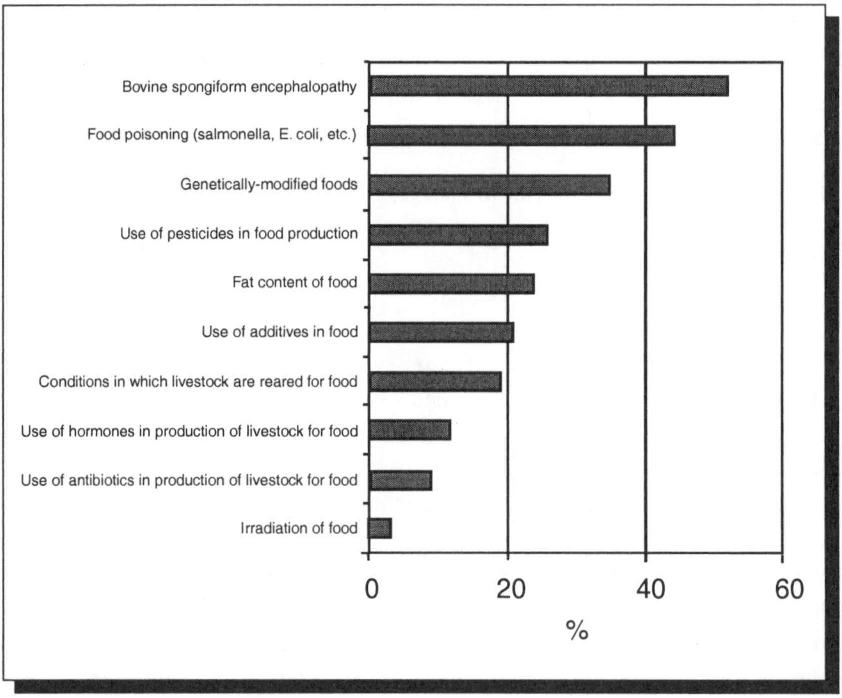

Source: Henson (forthcoming).

hand, whilst "scares" about the safety of food are an occasional occurrence in all countries, Europe has had more that its fair share in recent years. Second, there is evidence from analysis of newspaper content that media coverage of food issues, and emerging food safety issues in particular, is more prevalent in Europe than the United States. The risk perception literature refers to the social amplification of risk, whereby psychological, social, institutional, and cultural processes act to heighten or attenuate perceptions of risk and shape risk-related behaviour (Kasperson 1992; Kasperson *et al*. 1988). Media coverage of risk information is an integral part of the risk amplification process.

Figure 3 presents the results of a content analysis of selected newspapers in the UK and the US over the period 1995–98 (Kalaitzandonakes and Marks 1999). This analysis indicates that, while the level of coverage of agricultural biotechnology increased in both the UK and US over this period, in all years except 1995 coverage was significantly greater in the UK's *Daily Telegraph* than the US newspapers *USA Today* and *Washington Post*. Furthermore, over this period the con-

FIGURE 3: Coverage of Agricultural Biotechnology in Selected UK and US Newspapers, 1995–98

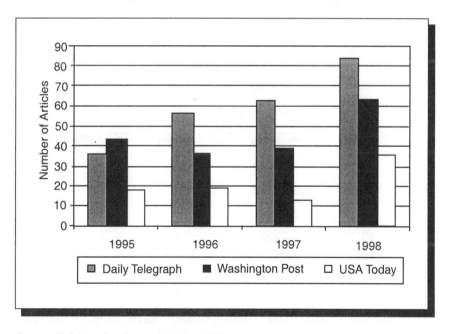

Source: Kalaitzandonakes and Marks (1999).

tent of media reporting of agricultural biotechnology in the UK was found to have become more negative, whereas in the US there had been no significant change.

A further issue influencing, and maybe also reflecting, consumer concerns about food safety in Europe is the lack of trust in public and private institutions as providers of information on food safety. For example, Figure 4 reports the results of a Eurobarometer survey undertaken for the European Commission across all EU member states in 1998 (INRA 1998). Specifically, it reports the proportion of respondents that consider various institutions to always provide truthful information about food safety. All institutions, both public and private, although with the one exception of consumer organizations, are judged to be unreliable sources of information — less then 30 percent of respondents considered that they always tell the truth about food safety.

The issue of trust in information on food safety is, however, a complex one (see Frewer and Shepherd 1994; Frewer *et al.* 1996; Hunt and Frewer 2001), involving, among other things, perceptions of an information source's credibility (Gutteling and Wiegman 1996). In turn, there are two dimensions to source credibility (Eagley, Wood and Chaiken 1978). First, "reporting bias," which refers to

FIGURE 4: Proportion of EU Consumers Considering that Persons or Organizations Always Tell the Truth about Food Safety, 1998

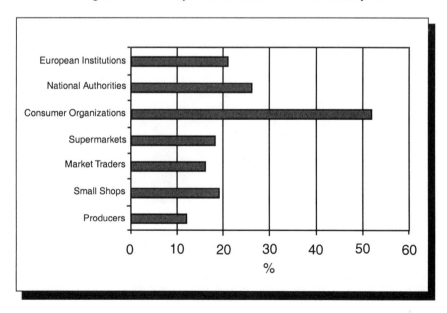

Source: INRA (1998).

the degree to which the information source can be expected to be truthful about a particular issue. Second, "knowledge bias," which refers to the perceived extent of expertise or knowledge of a source about an issue. The survey results in Figure 4 only address the first of these dimensions. However, given that respondents perceive such a high level of "reporting bias" for all institutions, with the exception of consumer organizations, they suggest that overall source credibility is low, regardless of whether these institutions are regarded as knowledgeable or not.

## PERCEPTIONS OF THE FOOD SYSTEM

Although consumer concerns about food safety are often discussed in isolation, in reality they are both embedded in, and reflect, more general perceptions of the performance of the food system. Thus, *ceteris paribus*, we might reasonably expect consumers to be more concerned about food safety if they perceived that the food system as a whole is performing badly in terms of the issues that are important to them. These issues might encompass the range of food products available, cost of food, behaviour of food companies, where food comes from, convenience, health and nutrition, and ethical issues associated with food production.

Figure 5 reports the results of a study that employs multi-item scaling to elicit the perceptions of UK and US consumers regarding the performance of the food system (Henson and Traill 2000). Data were gathered through a random postal survey of 1,000 food purchasers in the UK and US in 1998.[5] Performance of the food system was evaluated with respect to nine welfare constructs — elements of the food system that were identified as affecting the welfare of food consumers. For each construct, seven attitudinal statements were presented and respondents asked to indicate the extent to which they agreed with each on a seven-point Likert scale from "disagree strongly" at one extreme to "agree strongly" at the other. This enabled seven-point performance scores for each construct to be derived that could be aggregated to provide an overall measure of the perceived performance of the food system.[6]

The results of this study show surprising similarities between the perceptions of UK and US consumers. First, the performance scores of all welfare constructs were low, indicating that the food system as a whole was perceived to be performing relatively badly. Second, elements of the system that were perceived to be performing least well were ethical issues and behaviour of food companies. These results suggest that there may be little difference in the level and nature of concerns about the food system as a whole between European (or at least UK) and US consumers.

In both the UK and the US, safety was a construct perceived to be performing better than most others. Indeed, only convenience and choice were perceived to

FIGURE 5: Perceived Performance of Food System in the US and UK, 1998

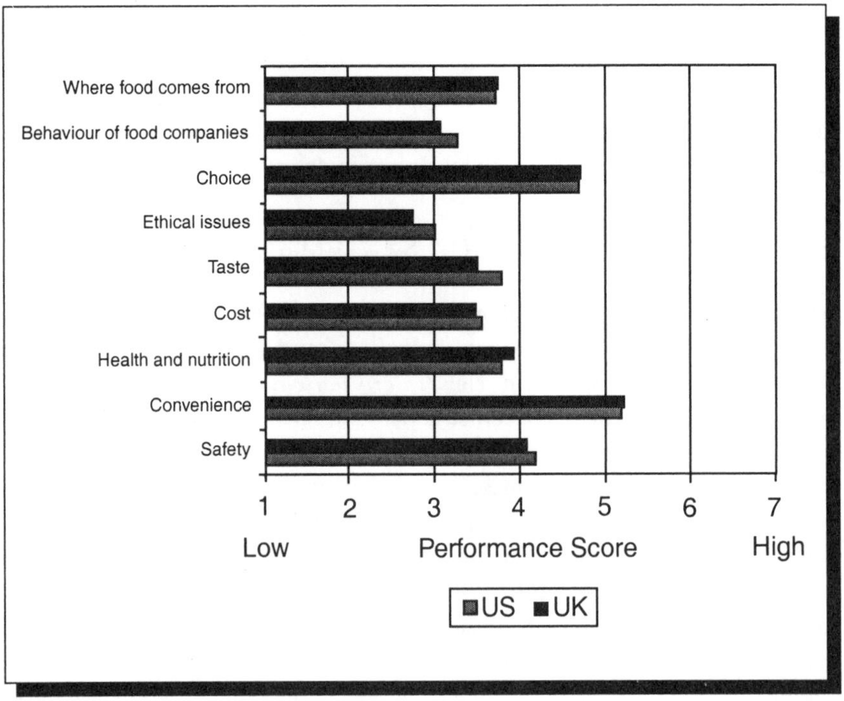

Note: Perceived Performance in UK and US not statistically significant at 5 percent level except for ethical issues and taste.

Source: Henson and Traill (2000).

be performing better. There was, however, a strong correlation (r=0.74) between the perceived performance of the safety construct and the aggregate performance of the food system as a whole. This was not true, however, with certain other constructs such as convenience and ethical issues. This suggests that concerns about food safety may indeed reflect the broader perceived performance of the food system as a whole.

## FOOD SAFETY CONCERNS AND RESPONSIBILITY

Consumer concerns about food safety are influenced by their perceptions of responsibility (Beck 1992, 1999; Leiss and Chociolko 1994). On the one hand, there is responsibility, on a day-to-day basis, for ensuring that food is safe to eat, or at least that the risks associated with food are within acceptable bounds. This raises questions about the role of, among others, government, food suppliers, and consumers themselves. On the other hand, there is a responsibility when food safety controls fail and consumers are exposed to a risk they perceive to be unacceptable.

Previous studies suggest that European consumers are sceptical about the efforts of public authorities and food suppliers to control food safety (see Becker 2000; Harper and Henson 2001). For example, Figure 6 reports the results of a representative sample survey of 500 food purchasers in the UK elicited by postal interview in 1998 (Henson, Griffith and Loader 1999).[7] Respondents were asked to rank a variety of agencies according to: (i) the degree to which they are responsible for ensuring the food they eat is as safe as possible; and (ii) the degree to which they actually take responsibility for ensuring that the food they eat is as safe as possible. Mean rank scores for each agency were then calculated, where one corresponds to most responsible and five to least responsible.

FIGURE 6: Consumer Perceptions in the UK of Responsibility for Ensuring the Food they Eat is as Safe as Possible, 1998

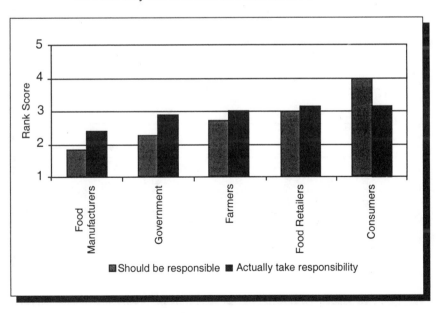

Source: Henson, Griffith and Loader (1999).

These results indicate that, while food consumers in the UK perceive food safety to be the primary responsibility of government and the food supply chain, in particular food manufacturers, they consider that much of the responsibility in practice is taken by consumers. For example, while food manufacturers are perceived to be primarily responsible for food safety, they are seen to actually take the least responsibility. In contrast, consumers are perceived to be least responsible for food safety, but to take most responsibility in practice. This contrasts with the United States, for example, where consumers, at least one survey suggests, see themselves to be primarily responsible for the safety of the food they eat (FMI 1998).

FIGURE 7: Perceived Risks and Benefits to Consumers of Food Technologies

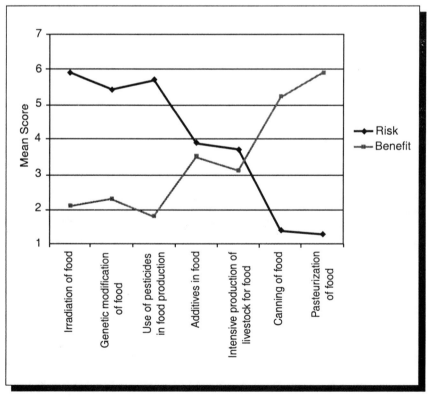

Source: Henson (forthcoming).

The acceptability of food-related risks and perceptions of responsibility are influenced by the trade-off between risks and benefits as perceived by consumers (Alhakami and Slovic 1994; Frewer *et al.* 1997; Leiss and Chociolko 1994). This assumes that consumers perceive risks and benefits as distinct elements and compare their relative magnitude and distribution when determining the acceptability of the risks they face.[8] Thus in the case of new food technologies, risks will tend to be more acceptable, and consumers will be willing to take more responsibility for the risks they face, where there is perceived to be a net benefit to them personally.

Figure 7 reports the results of a random postal survey of 1,000 food purchasers in the UK in 1999 (Henson forthcoming).[9] Respondents were presented with seven food technologies and asked to score the risks and benefits of each to them personally on a seven-point Likert scale from "minor risk/benefit" at one extreme to "major risk/benefit" at the other. For three technologies, namely irradiation, genetic modification, and use of pesticides, respondents considered the risks to far exceed the benefits, suggesting that the acceptability of these technologies is questionable. This contrasts with established technologies that are commonly accepted by consumers, namely canning and pasteurization, for which the benefits significantly exceeded the costs. There was a strong positive correlation between perceived risks associated with each technology and the respondents' perceptions of their knowledge of the technology.

## JUDGING SAFETY AT THE POINT OF PURCHASE

The ability of consumers to judge the safety of food products at the point of purchase and the manner in which they do so influences not only the magnitude of their food safety concerns, but also the impact on their purchase behaviour (Henson 2000). On the one hand, consumers who are able, or perceived that they are able, to judge the safety of food at the point of purchase may judge the risks they face to be controllable and voluntary, and thus more acceptable (Slovic 1992). On the other, the ability to judge the safety of food prior to purchase presents possibilities for averting behaviour on the part of the consumer, in both their purchase behaviour and food preparation practices.

The food safety characteristics of food cannot be observed directly by the consumer prior to purchase. In some cases, such as food poisoning, they experience characteristics that can only be observed post-consumption. In others, such as pesticide residues, they credence characteristics that cannot even be observed post-consumption (Henson and Traill 2000). As a consequence, consumers use other intrinsic and extrinsic product cues as indicators of the safety (and other aspects of quality) of the product (Grunert *et al.* 1997; Henson 2000). Indeed, through their own experiences and other sources of information, they tend to infer link-ages between observable product characteristics and the safety of the product. Examples include price, country of origin, brands and other quality marks, and product appearance.

Henson and Northen (2000) investigate the process by which food purchasers assess the safety of beef at the point of purchase. The ability of consumers to assess the safety of beef reflects their judgements of the efficacy of observable safety indicators, including price, knowledge of animal feed, organic, and brand/quality label. The efficacy of these indicators was assessed through a telephone survey (n=500) of food purchasers in six EU member states in 1998. Figure 8 reports the ranking of food safety indicators according to influence on perceived ability to judge the safety of beef at the point of purchase in Ireland, UK, Sweden, and Germany.[10] In all of these countries observed freshness and country of origin (with the exception of UK) were the most important determinants, whereas price and name of the producer were consistently among the least important.

FIGURE 8: Ranking of Safety Indicators in Terms of Ability to Judge Safety of Beef at Point of Purchase in Ireland, UK, Sweden and Germany

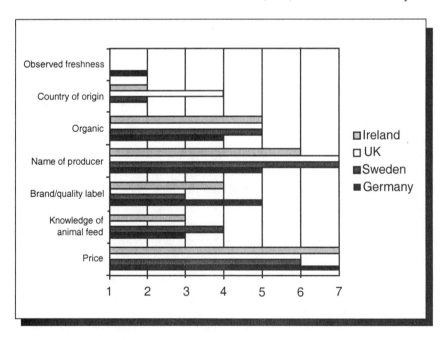

Source: Henson and Northen (2000).

## IMPACT ON PURCHASING BEHAVIOUR

The foregoing discussion has explored the degree to which consumers are concerned about food safety in Europe, with some comparison with the United States, and the nature of these concerns. The next key question is the extent to which these concerns are reflected in consumer purchasing behaviour. In many ways, however, this is a difficult issue to address, since reported purchase behaviour and purchase intention are notoriously unreliable measures of actual behaviour. For example, Figure 9 reports the results of a recent telephone survey of 500 food purchasers in each of five EU member states (UK, Ireland, Germany, France, and Italy) which explored consumer attitudes toward animal welfare (Harper and Henson 2001). When respondents were asked whether they tended to purchase products that were labelled as having being produced with higher standards of animal welfare than normal, in four countries over 50 percent claimed to do so. The actual market share of most animal welfare-related products in these countries is, however, typically less than 5 percent (Harper and Henson 1998). This

FIGURE 9: Consumers Tending to Select Products Labelled as Having Been Produced with Higher Standards of Animal Welfare than Normal

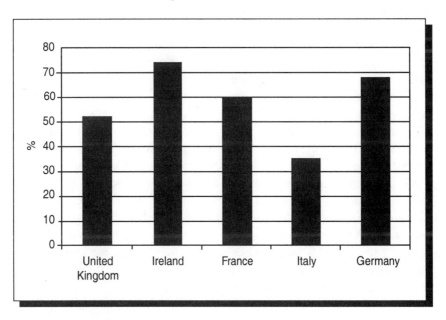

Source: Harper and Henson (2001).

suggests that, whilst respondents thought it desirable to purchase such products, or at least for others to believe that they did, this is unlikely to have been reflected in their actual purchase behaviour (Harper and Henson 2001).

It is evident, however, that food safety concerns can potentially have a significant impact on both intended and actual purchase behaviour. For example, a number of contingent valuation (Buzby, Ready and Shees 1995; Henson 1996; Henson, Griffith and Loader 1999; Ready, Buzby and Hu 1996; Weaver, Evans and Luloff 1992) and experimental market studies (see for example Fox *et al.* 1994; 1998; Hayes *et al.* 1995; Stefani and Henson 2001) indicate that consumers are willing to pay a price premium to secure higher levels of food safety. For example, Henson (1996) estimates that the implied "willingness to pay" of UK consumers to avoid a case of food poisoning in 1996 was £2,554 for eggs and £5,446 for chicken. Furthermore, demand modelling studies indicate that the provision of risk information can have a significant impact on the market demand for food products (see Brown and Schrader 1990; Flake and Patterson 1999; Henneberry, Piewthongngam and Qiang 1999; Smith, van Ravenswaay and Thompson 1988; van Ravenswaay and Hoehn 1991).

As well as specified improvements in food safety, consumers also value the implementation of government food safety controls, although the outcome of these controls in terms of levels of food safety may be uncertain. Henson, Griffith and Loader (1999) estimate the value placed on a government program of research and surveillance for chemical contaminants in food, which aims to minimize the risk that contaminants are found in food at levels that scientists judge to be unacceptable. Contingent valuation was employed with single-bounded dichotomous choice as the elicitation method and specified increases in weekly food expenditure as the payment vehicle. The survey was undertaken by telephone on a stratified sample of 550 food purchasers in 1998. A total of 71 percent of respondents were willing to pay to implement the program. The vast majority of the 29 percent zero bids were protest votes and excluded from the analysis. Figure 10 reports the distribution of responses to the offer price. Mean "willingness to pay" was £0.73/week, with a 95 percent confidence interval of £0.67 to £0.78.

The research reported above suggests a simple trade-off between money — in the form of product price — and the level of food safety. However, in making their product choices, consumers may simultaneously trade off food safety against price and a wide range of other product characteristics, such as taste and visual appeal (Baker 1999). Indeed, this can explain why, although consumers may be concerned about food safety, it is not immediately obvious that this is reflected in their purchase behaviour.

Henson and Azam (2001) employ conjoint analysis to explore the trade-off between food safety and other product characteristics in the choice of tomatoes.

FIGURE 10:  Proportion of Respondents Willing to Pay Offer Price to
           Implement Government Controls on Chemical Contaminants in
           Food in the UK

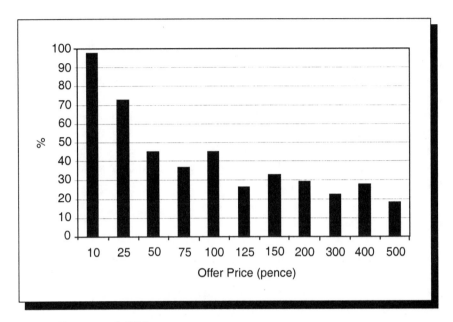

Source: Henson *et al.* 1999.

Consumers were presented with tomatoes that varied in the pesticide regime
employed in production (organic, reduced-pesticide use, and conventional production), agency certifying the method of production, price, amount of damage
and country of origin.[11] Each characteristic had three levels to prevent bias in the
estimated part-worths. Product scenarios were presented to respondents in the
form of a photograph and a written description of product characteristics.

The survey was undertaken through personal interviews with 500 food purchasers in a range of supermarkets in the Reading area of the UK. For each product scenario, respondents were asked to indicate their preference on an 11-point
numerical scale. Part-worths were subsequently estimated for each respondent
using a linear model with no interaction between variables. The average $R^2$ for the
sample as a whole was 0.92, indicating a good fit of the data to the hypothesized
model.

Overall, method of production was of greatest importance to respondents, accounting for around 48 percent of variation in the utility of respondents across the
range of each characteristic (Appendix 1). Price and damage were also important,

each accounting for around 20 percent of variation. Furthermore, organic production was highly valued by consumers, with a part-worth significantly greater than for any of the other product characteristics.

To explore the presence of distinct subgroups of consumers within the overall sample, cluster analysis was performed on the relative importance scores of each characteristic for each respondent. A total of four clusters were identified which could be meaningfully classified in terms of the relative importance of particular product characteristics (Appendix 2). The largest cluster, accounting for 46 percent of the sample, was classified as safety-conscious. In this cluster, the type of production method accounted for 78 percent of the variation in utility over the range of each characteristic. Other clusters were quality and price conscious, which are self-explanatory, and holistic consumers, for which no specific characteristic was of particular importance. These results are broadly similar to those of Baker's (1999) study in the United States.

## CONCLUSIONS

Given the history of "food scares" that have plagued Europe in recent years, there is a great deal of interest in consumer concerns about food safety. Within Europe, there is an interest in better understanding the concerns that consumers have and how these might be mitigated through risk communication. Outside Europe, there is a desire to learn from Europe's experiences. However, whilst there is now a large body of research on consumer concerns about food safety, as well as a more general literature on risk perceptions and risk-related behaviour, our understanding of these concerns and, in particular, their impact on behaviour remains limited.

It is widely assumed that European consumers are more sensitive about food safety issues than in other developed countries. However, after acknowledging the problems of comparing results, some of the research reported in this chapter suggests that consumers may be no less concerned about food safety in other countries, although different issues may come to the fore. There is clearly a need for more cross-cultural research to address this issue and explore the degree to which consumer concerns differ between economic, social, political, and cultural contexts.

This chapter has sought to provide an overview of the nature of consumer concerns about food safety, with a particular focus on research work undertaken in Europe and with the perspective of an economist as a starting point. In so doing, however, it has served to illustrate how traditional methods of economic analysis alone cannot answer many of the questions we have about consumer concerns and their impact on purchase behaviour. Rather, there is a need for economists to adopt other methods, which are widely applied in disciplines such as consumer research, and adapt them to their specific requirements. A good example is multi-item scaling, which has great potential in research of this kind, but of which many economists remain unfamiliar.

Whilst this chapter has hopefully thrown some light on the nature of consumer concerns about food safety in Europe (and in particular the UK), it has also highlighted some of the problems of consumer research in this area. Specific examples discussed include the elicitation and measurement of consumer concerns about food safety in general, and specific food safety issues in particular, along with the impact of concerns on purchase behaviour. Clearly, methods of analysis need to be chosen with care and results interpreted in the light of the methods that are employed. Furthermore, results must be communicated with care, especially to the policymaker who is looking for a one-word solution!

## NOTES

1.  Much economics-based research on food safety has, however, employed research methods that have their origin in consumer psychology and market research.
2.  The response rate was 36.2 percent in UK and 16.4 percent in United States.
3.  This is a typical property of prompted question of this type (see, e.g., Foddy 1993).
4.  This item preceded the prompted question reported in Figure 1, which came much later in the survey instrument.
5.  The response rate was 31.4 percent in UK and 9.4 percent in US.
6.  Data were also collected on the importance of each construct to the welfare derived from the food system. This enabled the performance scores of each construct to be weighted by importance in deriving an overall measure of performance of the food system.
7.  This formed part of a contingent valuation study that is reported more fully in the section on the impact of purchasing behaviour.
8.  There is evidence, however, that perceived risk and perceived benefit are inversely related (see Alkhakami and Slovic 1994; Frewer *et al.* 1997)
9.  The response rate was 36.2 percent.
10. This was derived on the basis of relative values of standardized estimates of model parameters.
11. A description of each production method was provided to respondents indicating, in particular, the risks and benefits associated with each.

## REFERENCES

Adu-Nyako, K. and A. Thompson. 1999. "Food Safety, Risk Perceptions and Behaviour of Consumers in the Southern Black Belt Region of the US." Paper presented at the annual meetings of the American Association of Agricultural Economics, Nashville, August.

Alkhami, A.S. and P. Slovic. 1994. "A Psychological Study of Attitudes," *Risk Analysis* 14:1085-96.

Baker, G.L. 1999. "Consumer Preferences for Food Safety Attributes in Fresh Apples: Market Segments, Consumer Characteristics and Marketing Opportunities," *Journal of Agricultural and Resource Economics* 24(1):80-91.

Beck, U. 1992. *Risk Society: Towards a New Modernity*. London: Sage Publications.

————— 1999. *World Risk Society*. Cambridge: Polity Press.

Becker, T. 2000. *Quality Policy and Consumer Behaviour in the European Union*. Kiel: Wissenschaftsverlag Vauk Kiel.

Bredahl, L., K.G. Grunert and L. Frewer. 1998. *Consumer Attitudes and Decision-Making with Regards to Genetically Engineered Food Products*. Aarhus: Centre for Market Surveillance, Research and Strategy for the Food Sector.

Brown, D.J. and L.F. Scrader. 1990. "Cholesterol Information and Shell Egg Consumption," *American Journal of Agricultural Economics* 72(4):548-55.

Buzby, J.C., R.C. Ready and J.R. Shees. 1995. "Contingent Valuation in Food Policy Analysis: A Case Study of a Pesticide Risk Reduction," *Journal of Agricultural and Applied Economics*, 27(2):613-25.

Eagley, A.H., W. Wood and S. Chaiken. 1978. "Causal Influences about Communications and their Effect on Opinion Change," *Journal of Personality and Social Psychology* 36:424-35.

Eom, Y.S. 1994. "Pesticide Residue Risk and Food Safety Valuation: A Random Utility Approach," *American Journal of Agricultural Economics* 76(4):760-72.

Fife-Schaw, C. and G. Rowe. 1996. "Public Perceptions of Everyday Food Hazards: A Psychometric Study," *Risk Analysis* 16(4):487-500.

Flake, O.L. and P.M. Patterson. 1999. "Health, Food Safety and Meat Demand." Paper presented at the annual meetings of the American Association of Agricultural Economics, Nashville.

Foddy, W. 1993. *Constructing Questions for Interviews and Questionnaires: Theory and Practice in Social Research*. Cambridge: Cambridge University Press.

Food Marketing Institute (FMI). 1998. *Consumer Attitudes and the Supermarkets 1998*. Washington, DC: Food Marketing Institute.

————— 1999. *Consumer Attitudes and the Supermarkets 1999*. Washington, DC: Food Marketing Institute.

————— 2000. *Consumer Attitudes and the Supermarkets 2000*. Washington, DC: Food Marketing Institute.

Fox, J.A., D.J. Hayes, J.B. Kliebenstein and J.F. Shogren. 1994. "Consumer Acceptability of Milk from Cows Treated with Bovine Somatropin," *Journal of Dairy Science* 77: 703-07.

Fox, J.A., J.F. Shogren, D.J. Hayes and J.B. Kliebenstein. 1998. "CVM-X: Calibrating Contingent Values with Experimental Auction Markets," *American Journal of Agricultural Economics* 80(3):455-65.

Frenzen, P.D., A. Majchrowicz, J.C. Buzby and B. Imhoff. 2000. *Consumer Acceptability of Irradiated Meat and Poultry Products*. Washington, DC: Economic Research Service, US Department of Agriculture.

Frewer, L.J. and R. Shepherd. 1994. "Attributing Information to Different Sources: Effects on the Perceived Qualities of Information on the Perceived Relevance of Information, and on Attitude Formation," *Public Understanding of Sciences* 3:3855-901.

Frewer, L.J., C. Howard and R. Shepherd. 1998. "Understanding Public Attitudes to Technology," *Journal of Risk Research* 1(3):221-35.

Frewer, L.J., C. Howard, D. Hedderley and R. Shepherd. 1996. "What Determines Trust in Information about Food-Related Risks? Underlying Psychological Constructs," *Risk Analysis* 16(4):473-86.

_____ 1997. "Consumer Attitudes Towards Different Food-Processing Technologies used in Cheese Production: The Influence of Consumer Benefit," *Food Quality and Preference* 8:271-80.

Glitsch, K. 2000. "Consumer Perceptions of Fresh Meat Quality: Cross-National Comparison," *British Food Journal* 102(3):177-94.

Grunert, K.G., H.H. Larsen, T.K. Madsen and A. Baadsgaard. 1997. *Market Orientation in Food and Agriculture.* Dordecht: Kluwer Academic Publishers.

Gutteling, J.M. and O. Wiegman. 1996. *Exploring Risk Communication.* Dordecht: Kluwer Academic Publishers.

Harper, G. and S.J. Henson. 1998. *Consumer Concerns about Animal Welfare and the Impact on Food Choice: A Review of the Literature.* Reading, UK: Department of Agricultural and Food Economics, The University of Reading.

_____ 2001. *Consumer Concerns about Animal Welfare and the Impact on Food Choice.* Reading, UK: Department of Agricultural and Food Economics, The University of Reading.

Hayes, D.J., J.F. Shogren, S.Y. Shin and J.B. Kliebenstein. 1995. "Valuing Food Safety in Experimental Auction Markets," *American Journal of Agricultural Economics* 77(1):40-53.

Henneberry, S.R., K. Piewthongngam and H. Qiang. 1999. "Consumer Food Safety Concerns and Fresh Produce Consumption," *Journal of Agricultural and Resource Economics* 24(1):98-113.

Henson, S.J. 1996. "Consumer Willingness to Pay for Reductions in the Risk of Food Poisoning in the UK," *Journal of Agricultural Economics* 47(3):403-20.

_____ 2000. "The Process of Food Quality Belief Formation from a Consumer Perspective," in *Quality Policy and Consumer Behaviour in the European Union,* ed. T. Becker. Kiel: Wissenschaftsverlag Vauk Kiel.

_____ Forthcoming. *Consumer Perceptions of Food Safety and the Impact on Food Choice: A Comparison between the UK and United States.* Reading, UK: Department of Agricultural and Food Economics, The University of Reading.

Henson, S.J. and J. Northen. 2000. "Consumer Assessment of the Safety of Beef at the Point of Purchase: A Pan-European Study," *Journal of Agricultural Economics* 51(1):90-105.

Henson, S.J. and M. Azam. 2001. *Trade-off between Food Safety and other Product Attributes in Food Choices: The Case of Pesticide Use in Tomato Production.* Reading, UK: Department of Agricultural and Food Economics, The University of Reading.

Henson, S.J. and W.B. Traill. 2000. "Measuring Perceived Performance of the Food System and Food-Related Welfare," *Journal of Agricultural Economics* 51(3):388-404.

Henson, S.J., B.H. Griffith and R.J. Loader. 1999. *Economic Evaluation of UK Policy on Chemical Contaminants in Food.* Reading, UK: Department of Agricultural and Food Economics, The University of Reading.

Hunt, S. and L.J. Frewer. 2001. "Trust in Sources of Information about Genetically Modified Food Risks in the UK," *British Food Journal* 103(1):44-62.

INRA. 1998. *La Sécurité des Produit Alimentaires.* Eurobaromètre 49. Paris: INRA.

Kalaitzandonakes, N. and L.A. Marks. 1999. "Public Opinion of AgBiotech in the US and UK: A Content Analysis Approach." Paper presented at the National Meeting of the American Association of Agricultural Economics, Nashville, August.

Kasperson, R.E. 1992. "The Social Amplification of Risk: Progress in Developing an Integrative Framework," in *Social Theories of Risk,* ed. S. Krimsky and D. Golding. Westport: Praeger.

Kasperson, R.E., O. Renn, P. Slovic, H.S. Brown, J. Emel, R. Goble, J.X. Kasperson and S. Ratick. 1988. "The Social Amplification of Risk: A Conceptual Framework," *Risk Analysis* 8(2):177-87.

Latouche, K., P. Rainelli and D. Vermersch. 1998. "Food Safety Issues and the BSE Scare: Some Lessons from the French Case," *Food Policy* 23(5):347-56.

Leiss, W. and C. Chociolko. 1994. *Risk and Responsibility.* Montreal and Kingston: McGill-Queen's University Press.

McGuirk, A.M., W.P. Preston and A. McCormick. 1990. "Toward the Development of Marketing Strategies for Food Safety Attributes," *Agribusiness: An International Journal* 6:297-308.

Ready, R.C., J.C. Buzby and D. Hu. 1996. "Differences between Continuous and Discrete Contingent Value Estimates," *Land Economics* 72(3):397-411.

Saba, A., S. Rosati and M. Vassallo. 2000. "Biotechnology in Agriculture: Perceived Risks, Benefits and Attitudes in Italy," *British Food Journal* 102(2):114-21.

Slovic, P. 1992. "Perception of Risk: Reflections on the Psychometric Paradigm," in *Social Theories of Risk*, ed. S. Krimsky and D. Golding. Westport: Praeger.

Smith, D. and P. Riethmuller. 2000. "Consumer Concerns about Food Safety in Australia and Japan," *British Food Journal* 102(11):838-55.

Smith, M.E., E.O. van Ravenswaay and S.R. Thompson. 1988. "Sales Loss Determination in Food Contamination Incidents: An Application to Milk Bans in Hawaii," *American Journal of Agricultural Economics* 70(4):513-20.

Sparks, P. and R. Shepherd. 1994. "Public Perceptions of the Potential Hazard Associated with Food Production and Food Consumption: An Empirical Study," *Risk Analysis* 14(5):799-805.

Stefani, G. and S.J. Henson. 2001. *Exploring the Value of Information about Food Safety Attributes.* Reading, UK: Department of Agricultural and Food Economics, The University of Reading.

Taylor Nelson. 2001. *Consumer Attitudes to Food Standards.* London: Taylor Nelson.

van Ravenswaay, E.O. and J.P. Hoehn. 1991. "The Impact of Health Risk Information on Food Demand: A Case Study of Alar and Apples," in *The Economics of Food Safety*, ed. J.A. Caswell. New York: Elsevier Science Publishers.

von Alvensleben, R. 1997. "Consumer Attitudes and Behaviour on the Meat Market in Germany," in *Problems of Meat Marketing: Seven Essays*, ed. R. von Alvensleben, S. von Cramon-Taubadel, A. Rohr and K. Schleyerbach. Kiel: University of Kiel.

Wang, Q., C. Halbrendt and N. Caron. 1996. "Differences between Retailer and Consumer Concerns about Seafood Safety: Evidence from Survey Data," in *The Economics of Reducing Health Risk from Food*, ed. J.A. Caswell. Bridgeport, CT: Food Marketing Policy Center, University of Connecticut.

Weaver, R.D., D.J. Evans and A.E. Luloff. 1992. "Pesticide Use in Tomato Production: Consumer Concerns and Willingness to Pay," *Agribusiness* 8(2):131-42.

Yeung, R.M.W. and J. Mossir. 2001. "Food Safety Risk: Consumer Perception and Purchase Behaviour," *British Food Journal* 103(3):170-86.

# APPENDIX 1
## Mean Part-worths and Relative Factor Importance from Conjoint Experiment Involving Trade-off between Tomatoes with Various Characteristics

| Characteristic | Part-Worth | Relative Importance % |
|---|---|---|
| Intercept: | 8.12 (4.11) | – |
| **Price** | | |
| £1.00/kg | –1.12 (1.83) | 20.1 |
| £1.50/kg | –1.92 (2.62) | |
| £2.00/Kg | –2.34 (3.52) | |
| **Damage** | | |
| None | 0.13 (0.38) | 19.8 |
| Minor | –0.23 (1.02) | |
| Major | –1.92 (2.94) | |
| **Production method** | | |
| Conventional | –3.11 (1.82) | 47.9 |
| Reduced-pesticide use | 0.14 (0.83) | |
| Organic | 2.67 (1.74) | |
| **Certification of method of production** | | |
| Government | –1.24 (1.45) | 5.9 |
| Retailer | 0.87 (1.12) | |
| Independent body | 2.06 (3.31) | |
| **Country of origin** | | |
| English | 0.74 (1.56) | 6.2 |
| Spanish | –0.32 (0.84) | |
| Dutch | 0.09 (0.32) | |

Note: Standard deviation given in parentheses.

Source: Henson and Azam (2001).

# APPENDIX 2
## Relative Importance of Characteristics in Conjoint Experiment Involving Trade-off between Tomatoes with Various Characteristics by Consumer Segment

| Characteristic | Consumer Segment | | | |
| --- | --- | --- | --- | --- |
| | Safety Conscious % | Quality Conscious % | Price Conscious % | Holistic Consumers % |
| Price | 5.2 | 11.4 | 56.9 | 32.9 |
| Damage | 4.9 | 50.1 | 18.6 | 22.5 |
| Production method | 78.2 | 29.4 | 12.2 | 24.2 |
| Certification | 9.1 | 2.1 | 3.1 | 6.1 |
| Country of origin | 2.6 | 7.0 | 9.2 | 14.3 |
| *Proportion of sample* | 46.2 | 22.5 | 20.1 | 11.2 |

Source: Henson and Azam (2001).

# 8

# Food Labels as a Method of Risk Communication at the Retail Level in the United Kingdom

*Catherine Humphries*

## INTRODUCTION

Consumers expect food to be safe, to meet their aspirations of quality and to provide good value for money. Increasingly, they also have ethical and practical concerns about the locations and the methods through which their food is produced. Often these concerns are based not on scientific, fact-based evidence but rather on perceptions and uncertainty fuelled by high profile, food-related health scares. These evolving consumer expectations present a communications challenge for retailers. In the United Kingdom, labelling has been used as a means of conveying information to consumers in an attempt to address their safety concerns. In developing an understanding of the communications challenge surrounding food safety in the UK, the following discussion considers the impact of the mad cow catastrophe and rising consumer concerns about genetically modified (GM) food on public opinion. The strategy of labelling is then considered as a mechanism through which retailers can communicate information to the public such as the risks associated with GM food.

## BRITISH BEEF AFTER MAD COWS

The British beef industry has faced particularly hard times in recent years. The industry was devastated by the contamination of the UK cattle herd with bovine spongiform encephalopathy (BSE). The situation was exacerbated by the belated official admission of a link between BSE in cattle and a comparable disease in humans, a new form of Creutzfeldt-Jakob disease (vCJD). The most likely cause of BSE was the practice of recycling animal protein in cattle feed, and it was transferred to humans through the consumption of parts of the central nervous systems of infected animals.

As a result of the outbreak of BSE, tens of thousands of cattle had to be destroyed. Furthermore, as consumers lost confidence in British beef, prices plummeted and fewer cattle were slaughtered for consumption leading to a rise in cattle stocks. Producers have faced a long and difficult struggle to regain these lost retail sales. Governments responded to the loss of confidence with a host of regulations, not the least of which, the European Union (EU) Beef Labelling Regulations, require all pieces of beef on sale to the public to be labelled. The labels must identify both the slaughterhouse and the cutting plant that handled each particular piece of beef. This requirement has proven to be an onerous task for butchery operations, and, while traceability is a key feature for dealing with any food scare, the value of the information to individual consumers is questionable.

The BSE crisis also heightened consumer concerns about animal welfare by drawing attention to some of the practices associated with intensive farming. More consumers now reject modern farming methods and express a desire to return to traditional farming practices. UK consumers have an idyllic vision of a countryside filled with cows, sheep, pigs, and hens grazing and roaming freely. They are generally concerned to find that the majority of animals reared for food are housed indoors for most, if not all, of their lives and may be subject to restraint. There is a particular abhorrence of the conditions in which young cattle are reared for veal, confined in veal crates and fed an unnatural diet to maintain the light colour generally associated with veal flesh. Governments responded with regulations that prohibit the use of veal crates in the UK. Consumers have also begun to protest the use of stall-and-tether systems in the intensive rearing of pigs. This increase in animal welfare regulation has led UK farmers to claim that they are unfairly placed to compete with farmers in other jurisdictions, because other European nations have been slower to regulate veal production. For instance, while stall-and-tether systems have now been phased out in the UK, the law continues to permit their use in the rest of Europe.

UK farmers have also criticized retailers for importing meat from Europe and other jurisdictions where, they allege, producers are not subjected to such stringent welfare standards as UK producers. However, this allegation is misplaced. Major British supermarkets, themselves under pressure from a Competition Commission review of their operations, have been responding to customers' preference for higher welfare standards. They have been instrumental in raising welfare standards throughout the world in order to provide their customers with food that meets the ethical and quality standards demanded by UK consumers. Nevertheless, not all animal products available in the UK will meet these higher standards and British products will continue to compete with less expensive and less regulated imports from abroad.

## GENETIC MODIFICATION

In recent years, the largest single consumer issue in the UK has been the introduction of genetically modified food products. Retailers, in particular, recognized the need to provide consumers with information about GM technology in order for it to be accepted by consumers. The critical issue, while educating consumers about the new technology, was to ensure respect for their right to choose. Labelling was viewed as the key to offering consumer choice, and an industry working group consisting of representatives from consumers as well as the retailing, manufacturing, and biotech industries was established to develop rules for product labelling.

The first genuine GM product to be marketed in the UK was a tomato paste sold by two supermarkets. Both supermarkets respected consumer choice by clearly labelling their product and offering it for sale alongside the regular tomato paste. Allegedly, the quality of the GM paste was superior to the standard product, but more tangibly, it offered consumers the benefit of being cheaper. The new GM tomato pastes were well received by consumers; sales went well and there was no evidence of consumer rejection.

The next GM products to reach the UK market were GM soy and GM maize. While these commodity crops are not used in large quantities in many foods, they are used in a whole range of foods in small quantities. Both GM products were introduced onto the world market through US producers, and there was no attempt to segregate GM crops or ingredients from their non-GM predecessors. As a result, it was unclear whether or not foods contained the GM varieties of soy or maize.

Thus, in the interest of progress, consumers were denied the ability to choose whether or not they would consume GM products. There was no discernible difference in the products they purchased and, although economic analysis could identify consumer benefits, they were not generally visible to individual buyers. Instead, the biotech companies (and possibly farmers) were seen as the primary beneficiaries of GM technology.

Consequentially, consumers began to ask why they should purchase GM food products. They questioned why their food was being manipulated, expressing concern that the process was unnatural. Although food safety has never been the primary concern with the products, many questioned the environmental safety of GM technologies and the long-term impact of interfering with nature. The primary concern has been with the lack of consumer choice. Consumers felt the technology was being forced on them with no personal gain and with potential, unquantifiable risks. They were concerned that they were being used as guinea pigs in a scientific experiment. Demands from consumers for the removal of GM ingredients from food products have become increasingly vocal. Thus, all the supermarket chains, and many branded food manufacturers, have adopted the non-GM stance initially taken by the Iceland supermarket chain. Essentially, all food marketed in the UK is intended to be from non-GM sources.

Pressure groups have recently targeted the use of GM ingredients in animal feed as an area for action. Despite the lack of a tangible food safety risk, consumers are beginning to demand animal products sourced from animals that have not been fed GM materials. All supermarkets are now looking to their suppliers to implement this strategy, placing a demand on world markets to supply adequate quantities of relevant feed ingredients from non-GM sources. In the longer term consumer demands could be expanded to include non-food products, particularly the more intimate products such as tampons and baby-care items.

## LABELLING

Mad cow disease and genetic modification are two areas in which labelling can provide the means to offer choice to consumers. They are becoming more discerning and there are a number of factors affecting food where, for a variety of reasons, consumers would like more information. Many of these information requests have been realized and voluntary codes have been drawn up to ensure that information is presented in a consistent format. The Food Standards Agency (FSA), recently established with a mandate to place consumers first, has already undertaken a review of labelling and adopted an ambitious action plan identifying a number of areas for future activity.

The FSA intends to lobby the European Union to require nutrition labelling on all foods and has commissioned research to define the ideal label content and format. Largely voluntary, nutrition labelling is currently controlled by EU regulations that prescribe the format to be used. It has been widely accepted that the current labelling process does not deliver information in a manner that is comprehensible, or useful, to consumers. A voluntary guideline has been introduced by industrial partners highlighting calories and fat as the nutrients of greatest interest to consumers. This policy recommends the use of "guideline daily amounts" as a means of placing the nutrition information of individual products into a context meaningful to consumers. The FSA would like to promote the use of "guideline daily amounts" in Europe in order to move away from the standardized, and largely incomprehensible, information required by the current directive.

The FSA is also encouraging the EU to enact legislative amendments that would remove the "all ingredient listing exemption" for major allergens, by deleting the "25 percent rule." For individuals with allergies the accidental consumption of an allergen can provoke anything from discomfort to a fatal reaction. Current labelling rules exempt certain ingredients from declaration in ingredient lists. For instance, the 25 percent rule allows any compound ingredient constituting less than 25 percent of the total food product to be listed by its name alone, declaring only certain additives used in its formulation. This creates a minefield for individuals with serious allergies who can suffer a reaction from molecular quantities of certain allergens. Although industry has adopted a guideline for the voluntary inclusion of the most common allergens in product labels, consumers are not protected from manufacturers who choose not to implement this guideline or foreign manufacturers who may be unaware of its existence.

The FSA is prompting the EU to enact legislation allowing for disease reduction claims to be included in product labels and to create an effective, practical system of verification and approval of claims at the EU level. UK law prohibits any claim that a food can prevent, treat or cure a disease, thus effectively preventing the linking of foods and diseases in product labels. With the current development of functional foods and advances in research demonstrating the link between human health and food components, it is important that this area be addressed in order to give consumers the opportunity to choose a healthy diet.

Additionally, the FSA would like the EU to clarify the rules governing the use of terms like "product of" and to implement origin labelling on a wider range of foods. In the UK it is the convention to ascribe the origin of a product to the location where it last underwent a major change. This might relate to where the product was manufactured but could also reflect where it was packaged or otherwise modified. Furthermore, the origin listed on the label bears no direct relationship to the origin of any of the ingredients. Research by the FSA has shown that this misleads consumers, who would like to know the source of major ingredients such as meat and dairy products. In part, this reflects consumers' interest in supporting local produce and local agriculture. Guidelines to address the confusion surrounding origin labelling have been published but are only applicable to producers in the UK. The FSA intends therefore to pursue the establishment of origin labelling guidelines in *Codex Alimentarius*.

Another area of concern to the FSA is the inclusion of alcoholic beverages in rules governing the listing of ingredients. Not only is there no legal requirement to indicate the ingredients in alcoholic drinks, it is actually illegal to do so in the case of wines. This situation undermines the individual's right to information about the products they are consuming and undermines their ability to make informed choices.

The FSA supports public requests for more information on food production systems. As has been demonstrated by genetic modification, consumers want to know how their food has been produced. In the past it was assumed that if the processes and procedures involved in production were safe, the only reason to inform consumers was in the event of a material change to the food itself. Thus, consumers were provided with information regarding the ultra high temperature (UHT) treatment of milk, a process that produces milk with a longer shelf-life than pasteurized milk. Now, consumers are demanding the right to be informed about all changes in production procedures, so that they can choose whether or not to accept such processes. Of particular interest to consumers is the use of chemicals, pesticides, and veterinary medicines in the production of food.

The FSA has set up a task force with a mandate to improve label clarity. The current regulations governing the presentation of information are very general, requiring some information to be grouped together (i.e., the product description, size, and its expiry date) and all label information to be easily understood, clearly legible, and indelible. Yet consumers complain that vital information is often hidden on the back of packages and in small print, rendering it illegible, particularly when shopping without spectacles! There is much that can be done to improve the clarity and accessibility of labelling information. The key issues are whether consumer concerns can be met through a prescribed label format and whether alternative sources of information will prove to be an acceptable means of relieving the congestion of information on labels. Such alternative sources could involve opportunities presented by new technology, as well as initiatives to provide additional, off-label information.

Finally, the FSA will address the need to improve information provided on food products that are sold loose or consumed in restaurants and other dining establishments. Currently, the amount of information provided to consumers purchasing unwrapped foods or eating in pubs, cafés, restaurants, and similar outlets is very limited. While consumers acknowledge these limits, their desire for information is offset, particularly when eating out, with a desire for nothing to interfere with their dining pleasure. However, certain information may be essential to some consumers, for example, those suffering allergies. Currently, food providers are required to be capable of informing customers whether the food products offered for sale contain genetically modified ingredients. It is likely that food providers will be required to provide additional information to consumers in future; however, further information will probably only be provided if requested by a customer.

Consumers do show a preference for UK produce and we are seeing a distinct trend toward local sourcing. Whereas consumer behaviour in the past has been largely driven by considerations of quality and price, for some customers intangible or ethical issues are a growing consideration in their product choices. As the following examples illustrate, product labelling can facilitate the desire to make informed choices.

In Scotland, a range of products have been developed under a single logo, "Scotland the Brand," which first and foremost, guarantees the product to be of Scottish origin. A national supermarket chain, located within the Cotswold area, has marketed "Cotswold Lamb." This, essentially niche product, has proven an enormous success for the company, demonstrating the extent to which customers value qualities other than price, in this case through choosing to support local farmers and demonstrating their sense of community.

At the national level, the National Farmers Union has adopted a logo, the "British Farm Standard," to embrace the best in British farming. For consumers, the presence of this logo identifies primary produce that has met the exacting standards prescribed by the National Farmers Union. An essential feature of this system is that agricultural produce must be grown according to established standards. The standard applied will vary depending on the type of product, such that horticultural products (i.e., fruit and vegetables) will be required to meet the standards of the Assured Produce Scheme, and meat products will be required to meet the standards of the Assured British Meat Scheme. To qualify, each of these schemes must utilize best agricultural practices. In practical terms this means that in the case of produce, the use of chemical inputs, pesticides, and fertilizers will be reduced through the implementation of integrated crop management. In the case of meat production, the adoption of best practices will limit the use of antibiotics and prohibit the recycling of waste animal products in animal feed. Another critical feature for qualification under the various schemes is independent verification to ensure the application of standards. The British Farm Standard logo replaces a multitude of individual logos developed by the individual produce sectors, supermarkets, and producer groups, thus presenting a consistent message to consumers regarding the quality and growing standards applied to their food.

Logos such as these can clearly be used to convey information to consumers regarding the products they choose to buy. In some cases the label is a guarantee of product source and in other cases provides information pertaining to production methods. Thus, labels enhance the ability of consumers to make informed choices.

## CONCLUSION

The advent of mad cow disease and the introduction of GM foods initiated widespread reaction on the part of consumers in the UK. Retail sales fell as traditional purchasing interests, such as price, gave way to new concerns such as food safety and ethical considerations regarding the processes involved in food production. Communicating information regarding risk to the public is a fundamental aspect of risk analysis. Thus, producers responded to consumer concerns through enhanced risk communication in the form of product labels designed to convey information. The provision of information (such as product origin, content, and production method) to consumers through labelling facilitated the ability of customers to make an informed choice about the products they were purchasing and contributed to the alleviation of consumer concerns regarding GM food products.

# 9

# Public Perceptions of Risk and the Regulatory Response in the United Kingdom

*Patricia Mann*

Consumers and institutions in the United Kingdom (UK) present an illuminating case study of the role that public perceptions and social movements play in influencing perceptions of and communications about risk. This chapter presents a discussion of public opinion regarding food safety issues and regulatory responsibilities, with particular attention to the UK context. The situation in the UK provides a useful insight into the complexity of factors that interact to shape public opinion and demonstrates the challenges involved in designing a regulatory framework against a backdrop of public mistrust resulting from perceptions of prior regulatory failures. As an illustration, I discuss the new UK Food Standards Agency (FSA) in light of its attempts to bring a greater degree of openness and enhanced risk communication into food safety policy. Its proponents hope that the FSA will provide the consumer with a single source of advice and information that will reduce the confusion and concerns arising from the multitude of conflicting opinions projected by the media.

## PUBLIC PERCEPTIONS

UK consumers are concerned and confused about a wide range of food-related issues. Their confidence has been battered by changes in conventional dietary wisdom and by a series of food-related health scares. Science cannot always give the public clear-cut advice or categorical assurances, but plenty of others will give their views — spread widely by the media and at times instantly and internationally via the Internet. The plethora of opinions and viewpoints has aggravated food safety concerns among consumers.

An Institute of Grocery Distribution study in August 2000 asked UK consumers about their food production concerns. When asked an open-ended question: "Thinking about the farm and factory, what are your main concerns about the way food is now produced," respondents mentioned up to five items (see Table 1). The top six were: hygiene (45 percent), animal welfare (40 percent), genetic modification (35 percent), pesticides (25 percent), bovine spongiform encephalopathy (BSE) (23 percent), and additives (14 percent).

Four of the top six (hygiene, pesticides, BSE, and additives) are clearly related to food safety. They have relatively well-accepted scientific evidence of their risks and the measures for handling them, as well as established risk-communications programs. The other two concerns (animal welfare and genetic modification) present a different challenge. The four food safety concerns relate to product safety but the other two are often characterized as issues involving the production method. Nearly half the respondents (49 percent) admitted to a poor understanding of how food is produced and 48 percent agreed that they were "not bothered about finding out." It is worth pointing out that while these survey results are from the UK, the ordinal ranking of risks is relatively similar to studies undertaken in the United States and Canada (Hoban 2001).

TABLE 1: Consumer Concerns about Food Production

*Question: "Thinking about the farm and the factory, what are your main concerns about the way food is now produced?"*

|  | *Percentage (%)* |
|---|---|
| Hygiene | 45 |
| Animal welfare | 40 |
| Genetic modification | 35 |
| Pesticides | 25 |
| BSE | 23 |
| Additives | 14 |
| Safety controls | 7 |
| Cost | 5 |
| Country of origin | 5 |
| Artificial insemination | 2 |
| Irradiation | 2 |
| Other | 1 |
| None | 4 |

Source: Institute of Grocery Distribution (2000).

TABLE 2: Consumer Concerns about Food Safety

|  | August 1998 | August 1999 | Percentage Change |
|---|---|---|---|
| Genetically modified food | 36 | 47 | +11 |
| Pesticides in fruit/vegetables | 37 | 36 | −1 |
| BSE/mad cow disease | 37 | 36 | −1 |
| Food poisoning in general | 36 | 32 | −4 |
| Cancer risk-related foods | 34 | 29 | −5 |
| E.coli | 35 | 28 | −7 |
| None of the above | 9 | 12 | +3 |

Source: Market Intelligence (1999).

Looking back at a series of studies undertaken by Market Intelligence in both 1998 and 1999, one can see some trends forming. In those surveys, consumers were prompted with options to identify up to five food issues about which they were most concerned. In that instance, the same six items emerging from the open-ended question a year or two later were of even greater concern. Perhaps most interesting is that while concerns were higher, for those with relatively effective risk-communication programs, the degree of concern was declining. Consumers only demonstrated increasing worry related to genetically modified (GM) foods, about which governments have not been able to effectively communicate.

That being said, one must remember that public opinion is by no means firm on food safety issues. The data on GM foods is inconclusive and can only lead to calls for more public information, and better two-way communication on the issue.

## FORMULATING A COMMUNICATIONS STRATEGY

In March 2000, a major qualitative research project was carried out in the UK on behalf of the new FSA. It has been set up to protect the public from health risks that may arise in connection with food consumption and otherwise to protect the food interests of consumers. The purpose of the research was to inform the development of a communications strategy to help position the new FSA. The approach of this research and the questions raised, as well as the findings, could usefully inform other discussions.

The quantitative and qualitative research was designed to test public attitudes to food safety, specifically to identify: general attitudes; levels of concern about food safety issues and priorities; confidence in or perceptions of threats in the food system; factors that influence food purchases (convenience, nutrition, cost, product information, ultimate consumer, etc.); knowledge, use and value of existing information sources; and confidence in levels of food safety across different sources (restaurants, fast-food outlets, etc.). The survey also tested perceptions of current management of food safety standards including perceived responsibility for ensuring standards, confidence levels in the government's ability to maintain standards, sources and requirements for information during scares, the attitude to the media during food scares and their perceived role, and awareness of the FSA. As well as establishing general benchmarks of attitudes, issues, information sources and relative confidence levels, the research was designed to identify public expectations of the proposed FSA, the public's understanding of its role, its impartiality, and its relevance in the current climate.

The research began by discussing food shopping and food choices in fairly general terms. A number of results became clear. When asked about responsibilities and priorities in shopping for food, most people talked first about convenience, price, and value. Next they stressed "quality," meaning primarily that food is fresh and appears to be hygienically prepared and presented. Difficulties mentioned by respondents were related to concerns about staying within a budget and, for parents, buying food that was both adequately nutritious and appealing to their children. On a more positive note, many respondents, especially older consumers, said food has improved: there is more of it and it is presented in a better manner.

"Quality" is most often associated with large supermarkets. There is trust in well-known, national names — large supermarket chains, producers of advertised, branded foods. They are seen as having the turnover and resources to ensure freshness and good management practices. Consumers expect them to react quickly to any complaints and to put things right. Smaller supermarkets, butchers, and market stalls were often seen as less reliable for hygiene and freshness, although the expertise of some smaller traders was recognized.

Few people have food safety issues on their mind when they go shopping. Respondents rarely mentioned food safety when asked about where they shopped or the food choices they made. Even when consumers mentioned food safety, there was confidence in the food they eat, attributed to trust in big retailers and food brands, belief that the UK has always managed its food well (and more hygienically than many other countries), a strong underlying confidence that the food industry is comprehensively regulated and the belief that "there is strength in numbers and no evidence that people in this country are getting ill, or are not well nourished."

When consumers were asked specifically about food safety, however, the responses were different. It is fairly clear that any discussion of the topic affects their views. They are perfectly comfortable talking about value or choices, but interpret food safety as bound up with larger, more intractable issues. Many instinctively feel it is, at the least, prudent to respond with a healthy scepticism to key underlying questions, such as: How much do we really know about how our food is produced and what it might contain? How confident are we that regulations are enforced? Do we think modern methods of mass food production are as safe as traditional methods? Do we trust food producers and retailers not to cut corners in the interests of profit? Do we think the science going into food production in the way of insecticides, pesticides, additives and so on has been tested thoroughly enough to be sure there will not be long-term ill effects? Do we trust our politicians to tell us the truth about food?

Any discussion of food scares simply provokes or sustains scepticism. In March 2000, three main concerns were at the forefront of *consumers'* food concerns. The BSE problem involved a catalogue of errors, including policies that allowed the feeding of cows with meat products, repeated government failures to communicate openly with the public, and lingering uncertainties about beef safety. Salmonella remained a concern. Finally, consumers are unclear about the issues related to GM foods and they have the feeling that something is amiss or not being well handled.

The FSA's research found that food safety scares have brought a number of underlying concerns to the surface. The public does not generally understand what is going on. Consumers feel they have to take things on trust but do not have much trust in authority, particularly in politicians. Consumers' feelings of uncertainty and powerlessness are increased by distrust of most media coverage. After consumers in the survey were engaged in the discussion on safety and scares, basic confidence in the goodness of food diminished and worries increased. Prominent worries ranged widely, including poultry-raising conditions; feed given to livestock (pigs, lambs, cows); effects of pesticides/chemicals on fruit and vegetables; additives in prepared food; and preparation and distribution conditions. Consumers questioned whether what they were eating could contribute to trends in diseases such as cancer and asthma and whether imported food products conformed to UK standards.

Newspaper stories were seen as typically sensationalist and brief, alarming without informing. Much media coverage was said to undermine trust and exacerbate public cynicism about the handling of food issues, by criticizing governments and by publicizing conflicting analysis and advice. Ultimately, the coverage gives the impression that experts cannot agree on the risks or the remedies. Consumers' distrust of much media coverage had given rise to feelings of uncertainty and powerlessness.

Consumer perceptions reveal some very real contradictions, which complicate public policy and risk communications. For instance, many consumers believe that the greater availability of organically produced foods "vindicates the need to be wary of modern food production." Furthermore, while big retailers and established brands are widely trusted, many consumer anxieties are seen to flow from the industrialization of farming and the commercialization of food distribution systems.

There is a widespread view that although adequate regulation may well be in place, there are very real problems with enforcement, especially for imported foodstuffs. Many feel that market forces can be relied upon to discipline both shops and restaurants effectively, but question whether it could ever be possible for governments to investigate and control the whole labyrinthine farm-to-fork chain. Nevertheless, they continue to look to governments for answers and direction.

In short, the FSA research ascertained that the public wants someone to blame for food scares and is quick to think that should be politicians. They want someone to be unambiguously responsible for and devoted to ensuring a safe food system. Although the new Food Standards Agency risks being seen as the creation of politicians, it was regarded very positively because of its dedicated purpose, authority, power and independence, not least from government departments.

## THE PUBLIC RESPONSE: THE NEW FOOD STANDARDS AGENCY

Created in 2000 as part of the UK government's response to concerns about food safety, the new FSA's primary objective is "to protect public health from risks which may arise in connection with the consumption of food (including risks caused by the way in which it is produced or supplied) and otherwise to protect the interests of consumers in relation to food" (FSA 2000). The FSA has adopted three core values to guide its work: to place the consumer first; to be open and accessible; and to be an independent voice. The FSA intends to mitigate some of the existing confusion and apprehension by providing independent advice and information to consumers regarding food safety concerns. Its credibility is enhanced by the appointment of Professor Sir John Krebs as its first chairman, described by the Oxford University Social Issues Research Centre as "a rational scientist with no political, commercial or ideological axe to grind."

In order to establish the FSA's openness, a series of public consultations were held on the FSA's working principles. The key change is that all discussions now take place openly. A code of practice on openness has been developed, making public disclosure of advice and information the norm. Openness and transparency mean that board meetings are held in public. The new agency uses the Internet as a means of disseminating information.[1]

The new agency has also taken a cue from the findings of the BSE Inquiry, published in October 2000. The report stated that at the heart of the BSE situation was the question of how to handle a hazard, particularly one in animals that was not known to occur in humans. The measures taken by the government — while often sensible — may not have been timely or adequately implemented or enforced. The government believed that the risk to humans was remote and therefore it was mainly concerned with preventing an alarmist over-reaction.

The FSA has published its own approach to risk. It is adopting an explicit precautionary approach, stating that it will not always wait for proof of a potential hazard to take action or issue advice. Instead, the agency will decide and act in proportion to its best judgement of the risks, including an assessment of the costs and benefits of the proposed action. Most importantly, the agency states that it will act quickly when necessary but will try to do so in a consistent and systematic manner. As for communications, the FSA aims to give everyone the information needed to make informed choices for themselves. While the needs and concerns of consumers are paramount, all those affected will be consulted. Finally, the FSA confirmed that it will learn from the experience of others at home and abroad, which will doubtless include the *Codex Alimentarius* experience of hazard assessment and risk management.

This new risk framework sounds good, but has yet to be truly tested. One major challenge is that consumer perceptions of risk differ from those of experts. Some argue that scientists and regulators tend to focus on measurable, quantifiable attributes whereas consumers give more weight to fairness, controllability, and "factors other than science," which include familiarity.

The challenge is for the FSA to provide accurate, proportional, and useful information. This would be a major challenge even if the FSA was the only actor in the field. But a European Union Food Safety Authority is planned to be operational by 2003 to collect and analyze information, issue scientific warnings and advice, and pre-empt or respond to crises. Representatives of European consumers are working with the Commission to make their needs known. The European Organization of Consumers Unions (BEUC) has called for farm-to-fork traceability, equivalent standards across all products, consistent definitions, restrictions on fortification and a nutrition action plan. The FSA is pressing the European Commission for new food labelling rules.

Others have plans to communicate with consumers about the food they eat. The UK Consumers' Association, for example, has launched its Fit to Eat Campaign with a series of magazine reports investigating the food chain, from plough to plate. In particular, it is seeking a new approach — as yet unspecified — to the assessment, management, and communication of food risks. The first report, in September 2000, was on food poisoning. Food poisoning was also the focus in 2000 of the Food and Drink Federation's National Food Safety Week, in cooperation with environmental health and trading standards organizations and government departments for Agriculture, Health and Education. In addition to leaflets on cleanliness, temperature, and storage, the National Food Safety Week's poster competition presented clear messages to schools and via school children into homes.

As an independent body that places consumers first, the FSA is intended to alleviate public concerns over food safety by providing individuals with the information and advice they need to make informed choices about the food they consume. The FSA will also act in an enforcement and monitoring capacity as well as promoting meaningful labelling in order to protect consumer health and promote informed decision-making. It is hoped that the FSA will mitigate public concerns regarding food safety by reducing the confusion generated by the multitude of opinions and perspectives projected by the media and other organizations concerned with food safety.

## CONCLUSION

It is too early to tell whether the new Food Standards Agency in the UK will help reduce or will exacerbate consumer fears about their food. Its success will depend on its ability to convince the public that it can adequately provide for their safety. This will not be a simple task given the contradictory nature of public opinion. Critical to the FSA's success will be the development of an effective communications strategy that enhances public understanding and allays concerns fuelled by an alarmist media that is projecting an inconsistent and confusing message. As research has indicated, food-related health scares are likely to undermine consumer confidence and focus attention on health-related concerns rather than the otherwise predominant consumer concerns of price and quality. An effective communications strategy will need to anticipate, or at the least respond in a timely manner, to food-related health scares in order not to undermine public trust that their food safety concerns are being properly addressed by the authorities.

### NOTE

1.   A statement on the FSA's approach to risk is available at <www.foodstandards.gov.uk/consultations.htm>.

REFERENCES

Food Standards Agency (FSA). 2000. *Qualitative Research to Explore Public Attitudes to Food Safety*. London: FSA.

Hoban, T.J. 2001. "Public Perceptions of Transgenic Plants," in *Transgenic Plants*, ed. G.G. Khachatourians, A. McHughen, W.-K. Nip, R. Scorza and Y.H. Hui. New York: Marcel Dekker.

Institute of Grocery Distribution. 2000. *Consumer Concerns*. London: Institute of Grocery Distribution.

Market Intelligence. 1999. *Consumer Attitudes Towards GM Foods*.

Phillips, Lord N., J. Bridgeman and M. Ferguson-Smith (Committee). 2000. *The BSE Inquiry: Return to an Order of the Honourable the House of Commons*, Report to Minister of Agriculture, Fisheries and Food, the Secretary of State for Health and Secretaries of State for Scotland, Wales and Northern Ireland, 16 vols. London: Her Majesty's Stationery Office.

# 10

# Risk Management and Communication: Enhancing Consumer Confidence

*Douglas A. Powell, Katija Blaine, Amber Leudtke, Shane Morris and Jeff Wilson*

## INTRODUCTION

There is a cacophony of food-related risks discussed in western media and society on a daily basis. Mad cow disease, pesticides, dangerous micro-organisms, hormones, animal welfare, and genetic engineering are prominently debated and then form the basis of consumer unease with respect to food safety. The potential for stigmatization of food is enormous, consequently, it is essential for risk managers to show that they are successfully minimizing a particular risk. Those responsible must be able to effectively communicate their efforts to consumers and must be able to prove they are actually reducing levels of risk. This chapter will introduce current risk management theory and present two examples of successful risk management and communication efforts on the part of producers.

## RISK MANAGEMENT

Food safety risks can be viewed, individually or collectively, as problems, and in some cases, opportunities, that can be evaluated using a risk-analysis approach. One such approach was advocated in a 1996 report by the US National Academy of Sciences' National Research Council Committee on Risk Characterization, which urged risk assessors to expand risk characterization beyond the current practice of translating the results of a risk analysis into non-technical terms. The report reframed risk characterization from an activity that happens at the end of the risk assessment process, as many individuals and agencies understand it, to a continuous, back-and-forth dialogue between risk assessors and stakeholders that is based on an iterative, analytic-deliberative process that allows the problem to be properly formulated.

Similarly, the US Presidential/Congressional Commission on Risk Assessment and Risk Management (1997) developed an integrative framework to help all types of risk managers — government officials, private sector businesses, and individual members of the public — make good risk management decisions. The framework has six stages: a definition of the problem and its context; an analysis of the risks associated with the problem in context; an examination of the options for addressing the risks; a decision about which options to implement, implementation of those options; and finally, an evaluation of the action's results (Figure 1). Of particular importance from a risk management and communication perspective is that the framework is conducted in collaboration with stakeholders and in an iterative manner.

FIGURE 1: The Risk Management Cycle

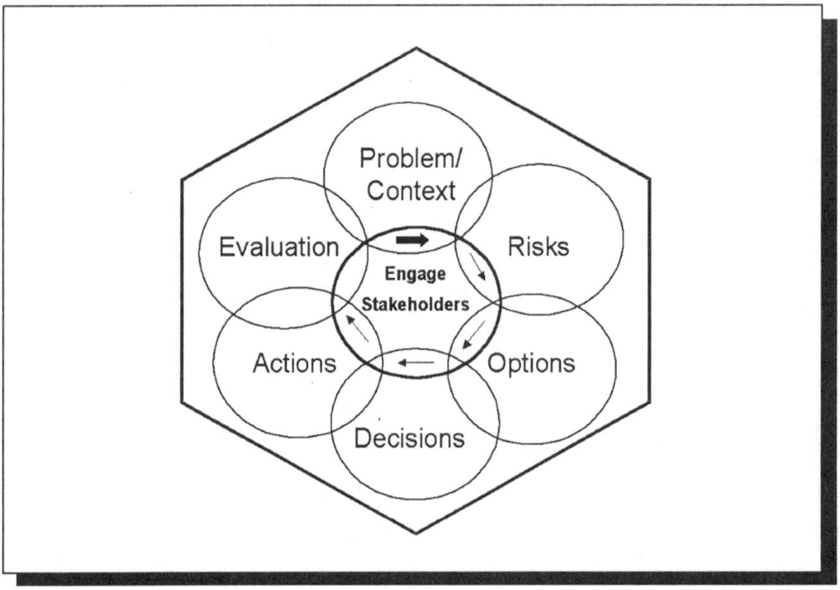

Source: US Presidential/Congressional Commission on Risk Assessment and Risk Management (1997).

The current state of risk management and communication research suggests that those responsible for food safety risk management must be seen to be reducing, mitigating, or minimizing a particular risk. Those responsible must be able to effectively communicate their efforts and be able to prove they are actually reducing levels of risk. Stigma is a powerful shortcut that consumers may use to evaluate food-borne risks. Gregory, Slovic and Flynn (1995) have identified factors that can lead to the formation of stigmata: if the source is a hazard; if a standard of what is right and natural is violated or overturned; if the impacts are perceived to be inequitably distributed across groups; if the possible outcomes are unbounded (scientific uncertainty); and if management of the hazard is brought into question.

The potential for stigmatization of food is enormous. Well-publicized outbreaks of food-borne pathogens and the furor over agricultural biotechnology are but two current examples of the interactions between science, policy, and public perception. The components for managing the stigma associated with any food-safety issue involve the following factors: effective and rapid surveillance systems; effective communication about the nature of risk; a credible, open, and responsive regulatory system; demonstrable efforts to reduce levels of uncertainty and risk; and evidence that actions match words.

Appropriate risk management strategies, such as on-farm food safety programs, as well as an honest discussion with consumers about agricultural technologies using a demonstration farm and market, are tools to enhance consumer confidence while producing scientifically safe food. They are also important and transparent mechanisms to collect data to validate arguments presented during international trade discussions. This chapter presents two case studies of food-safety risk management and communications as examples of effective risk management and risk communication.

## CASE STUDY 1: ON-FARM FOOD SAFETY PROGRAM

Outbreaks of food-borne illness associated with fresh fruits and vegetables have increased in recent years (Tauxe 1997). Consequently, methods of growing, handling, processing, packaging, and distributing fresh produce are receiving increasing attention in terms of identifying and minimizing microbiological hazards. The North American produce industry is focused on developing and implementing programs aimed at minimizing food-borne disease and illness. These Hazard Analysis Critical Control Point (HACCP)-based systems help to reduce the potential for microbial contamination along the entire production/distribution process, and subsequently limit the adverse effects of recalls, loss of sales or even food scares, all of which can ultimately lead to reduced sales and profits. Each food producer, from farm-to-fork, has a responsibility to ensure the safety and quality of their products. This preventive, proactive role adopted by the produce industry is a safeguard for the health and safety of the consumer (Hedberg, MacDonald and Osterholm 1994; Rushing, Angulo and Beuchat 1996; Tauxe 1997; Zhuang, Beuchat and Angulo 1995).

In 1998, the Ontario Greenhouse Vegetable Growers, which represents the approximately 220 growers of greenhouse tomatoes and cucumbers in Ontario, began working with the University of Guelph to design, implement, and analyze a set of practical and comprehensive guidelines that could be utilized by Ontario greenhouse vegetable growers in their daily operations. The On-Farm Food Safety program was based on a systematic approach to identify potential sources of microbiological hazards associated with fresh greenhouse produce. The objectives were validated through the use of microbiological testing, food-risk perception surveys, and on-site visits.

A thorough review of previous outbreaks, potential problem areas, and guidelines issued by government and other producer groups led to the development of a draft on-farm food safety manual, which contained general guidelines for the safe production and distribution of greenhouse vegetables. These guidelines were based on earlier studies which had identified possible sources of microbial contamination, more effective approaches to disinfection, and safety control measures to reduce the likelihood of food-borne outbreaks, especially salmonellosis, due to the consumption of fresh tomatoes and cucumbers.

A checklist format was developed highlighting parameters that required control, including water quality, hygienic practices, and handling and distribution procedures. A greenhouse product is grown under controlled conditions, harvested by hand, washed in a dump tank (tomatoes only), sorted by size and colour, and packed by hand (a series of pictures and videos illustrating greenhouse production can be found at: www.plant.uoguelph.ca/safefood).

The draft manual was presented at several meetings arranged with growers, as well as the annual meeting of the Ontario Greenhouse Vegetables Growers, and feedback was obtained from managers, farmers, and processors. The guidance document was subsequently revised to incorporate the best scientific advice with the goal of practical implementation in the greenhouse, packing shed, and distribution sectors. Further, baseline microbiological analyses, focusing on dump tank water quality and end product, were conducted.

The on-farm food safety manuals were distributed to all 220 growers in June 1999. Subsequently, a full-time employee was hired to implement the on-farm food safety program and to conduct site visits. During these visits, the importance of an on-farm food safety program was discussed, feedback from growers and processors was obtained, documentation reviewed, and water and end-product samples collected. It is important to note that samples were collected only from packing sheds, after tomatoes had been washed, sorted, and packed, and that numerous growers supplied the packing sheds.

Of the 37 produce samples collected in year one, all were negative for salmonella and E. coli, while four of the 37 had high levels of total coliforms. Of the six dump tank tests, three contained water with high total coliforms. Water supplies were also tested, of which 43 of 205 had high total coliforms, including five of 119 municipal systems. Results were communicated and preventative steps, such as adding additional chlorine to dump tank water, were recommended.

The first year represented a significant undertaking in terms of familiarizing growers and processors with the importance of food safety. In the second year, beginning in June 2000, sample collection and farm visitations were modified to improve the program based on grower feedback.

In addition to microbiological testing, three food-risk perception surveys were designed and distributed to Ontario Greenhouse Vegetable Growers' members in the spring of 1998, the fall of 1999, and the fall of 2000. Figure 2 shows respondent's perceptions of the greatest food risks in an unprompted question, while Figure 3 shows responses to a prompted question, such as, "Here is a list of possible risks associated with food." Figure 4 shows the programs that respondents had implemented to enhance the safety of their produce. Taken together, these results demonstrate an increased awareness of microbial food-safety risks among respondents and an increase in awareness of the importance of basic sanitation and hygiene.

FIGURE 2: What do you feel are the greatest threats to the food that you eat?

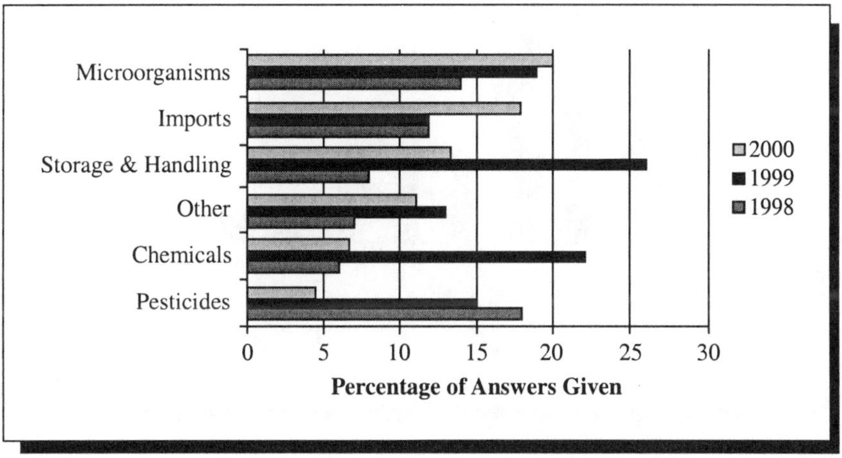

Note:  n = 53, 1998; n = 68, 1999; n = 45, 2000.

FIGURE 3: Rank the hazard of the following food items questions

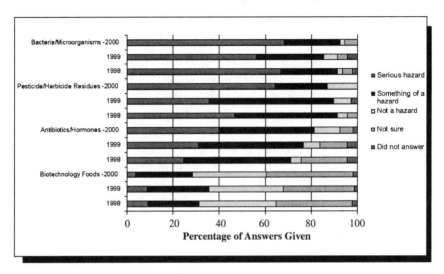

Note:  n = 53, 1998; n = 68, 1999; n = 45, 2000.

FIGURE 4: What programs or steps have you implemented that you feel would improve the safety of your products?

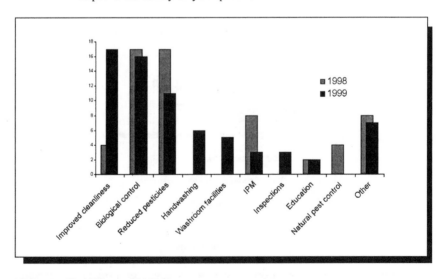

Note:  n = 68, 1999; n = 53, 1998.
Respondents may have indicated more than one area.

## CASE STUDY 2: DEMONSTRATION FARM AND MARKET

*Background*

Public debate regarding genetically engineered foods has increased significantly — internationally through the latter part of 1998 and specifically in Canada in the fall of 1999.

Genetically engineered (GE) Bt crops were designed to resist insect damage and reduce the need for chemical insecticides. Bacillus thuringiensis, or Bt, is a soil bacterium that produces a protein that is toxic to certain species of insects. While agricultural biotechnology and more specifically Bt crops have been extensively evaluated in a scientific setting, they have rarely been evaluated in the purchasing environment where consumers and primary producers actually make buying decisions.

Canadian surveys have found that most Canadians have heard of biotechnology and many of its applications (Angus Reid Group Inc. 1995). Awareness is greatest with respect to biotechnology's medical applications (Decima Research 1993; Einsiedel 1997), and this awareness is increasing (Einsiedel 2000). Canadians have indicated general concern over the use of biotechnology and genetic engineering in particular (Angus Reid Group Inc. 1995; Decima Research 1993; Einsiedel 2000). However, research indicates that consumer support for the products of genetic engineering are unlikely to be determined by attitudes to the technology overall; rather, acceptance may be based on recognition of tangible benefits of specific applications (Decima Research 1993; Frewer, Shepherd and Sparks 1994; Hoban 1997; Sheehy 1996; Zimmerman *et al.* 1994). It is evident that Canadian consumers are concerned about the safety of the food they eat and although bacteria and pesticides have been identified as the top concerns (CFIA 1998), concerns over genetically altered foods may increase as media coverage and consumer awareness increases.

Choice and honesty are fundamental values for consumers (Powell 2000). With respect to genetically engineered foods, the primary issues of concern identified by consumers through surveys and interviews include human and environmental health and safety, socio-economic, and ethical concerns (Decima Research 1993; Hagedorn and Allender-Hagedorn 1997; Sheehy 1996). Open, honest communication with consumers is essential to address these concerns. One, albeit limited, form of communication is product labelling. However, such labelling must be meaningful and understandable, which is a challenge given the low existing knowledge levels (Phillips and Isaac 1998).

Public perception research also indicates a high level of concern among the public over the use of pesticides in agriculture (Peterson 2000), specifically concerning human and environmental health impacts. Although some studies indicate that pesticides are still a greater concern than genetic engineering, biotechnology is less well understood than pesticide use (ibid.). This concern has led to the development of newer pesticides that break down faster in the environment with reduced human health risk profiles.

## Project Description

The purpose of this project was to examine the comparative trade-offs involved in producing Bt sweet corn and potatoes when compared with conventional varieties, ranging from the primary producer through to the consumer in a true farm-to-fork comparative evaluation.

Beginning 17 May 2000, three test plots in Orton, Ontario were planted with both genetically engineered Bt sweet corn and conventional sweet corn. The plantings were designed to simulate commercial production conditions. Each eight-acre plot was planted with four different varieties of corn in 20-row blocks, in random order.

For each planting, two spray lanes were left between every ten rows. The corn was planted in three different plantings for two reasons: first, to simulate commercial production and provide the continuity of supply that the fresh sweet corn market demands; and second, to determine the value of the Bt technology throughout the growing season as opposed to a "window" of critical pressure.

A press conference was held on 14 May to announce the research and a press release was distributed to underscore the public nature of the project. Throughout the summer, the project was featured on television and radio, and in the print media. Three press releases were issued in total: one announcing the project; one describing the opening of a 3km walking trail for curious visitors to the farm; and one highlighting the harvest and some of the findings from the farm production end.

Ten days later, 27 May, a public meeting was held at Birkbank Farms to inform local residents and neighbours about the project and to address any concerns among members of the community. Prior to this, letters were hand delivered to all the immediate neighbours and an advertisement was placed in the local newspaper, which explained the project and invited community members to the information session.

In order to establish and maintain open communication with the customers of Birkbank Farms, posters and bookmarks were displayed in the market, which described the project. Pamphlets containing background information on Bt sweet corn, as well as information about how to find out more about Bt sweet corn and genetically engineered crops, were also available in the market area.

A Web page was developed as a tool for communicating relevant aspects of the project and its progress with the media, industry, research community, and interested members of the general public. The Web site contained information on all the research protocols, background information on Bt sweet corn and other Bt crops, a weekly video and pictorial update on the research and the sweet corn, and any press releases and relevant news items.[1]

## Planting and Harvesting

A contact herbicide bromoxynil (product name Pardner) was used on all three plantings on 2 July and fertilizer was applied at 79lbs/ac on 16 July. The conventional corn on all the plantings were sprayed with the fungicide chlorothalonil (under the product name Bravo) for rust on 15 August. Professional scouts checked the corn plots two times per week looking for European corn borer (ECB) as well as aphids and beneficial insects such as lady beetles. The first and second plantings of corn were sprayed for ECB and ear worm on 14 August with carbofuran on half the conventional plants in both plots and Dipel DF (Bt spray) on the other half as recommended by scouts.

The first planting was ready for harvest on 28 August. The genetically engineered Bt corn required no sprays aside from the herbicide spray and fertilizer treatments. Half of the conventional corn was sprayed with five applications of Bt foliar spray and half was sprayed with four applications of carbofuran.

In the second planting, the genetically engineered Bt sweet corn also did not require any pesticide sprays, while the conventional sweet corn was sprayed with two applications of carbofuran and one application of cypermethrin (a pyrethroid under the product name Cymbush).

The third planting had the same herbicide and fertilizer applications as the other two plantings. However, the conventional sweet corn was sprayed with two applications of cypermethrin and one application of carbofuran.

The Bt sweet corn required neither insecticide nor fungicide applications, unlike the conventional sweet corn. The ECB and ear worm pressure were greatest in the first planting of corn and the cool, damp weather conditions made control of ECB and ear worm difficult, resulting in the heavy applications of carbofuran. The second and third plantings did not suffer the pest pressure of the first planting and insects were effectively controlled using cypermethrin in place of carbofuran in one or two instances. The scouts also noted that pesticide sprays killed nontarget insects.

According to the professional scouts, the regular corn suffered extensively greater worm damage than did the Bt sweet corn. Worm damage was found in 10 to 20 percent of the regular corn and only 1 percent of the Bt sweet corn. In the market, eight to ten cobs out of 100 of the conventional were thrown out daily due to worm damage whereas almost no Bt corn was thrown out for this reason. The Bt sweet corn was also found to have a longer shelf life. The last harvest of sweet corn was on 26 September 2000. The corn was then kept in cold storage and sold as needed. The regular sweet corn lasted until 6 October after which the quality was so poor it could not be sold. The Bt sweet corn was saleable until the week of 20 October.

The sweet corn was harvested by hand, segregated and packed onto labelled skids, and stored in cold storage until sale. Picking and packing was found to be less time-consuming for the Bt than for the conventional sweet corn due to significantly reduced worm damage on the Bt corn — so much so that the packers did not have to fully examine the ear of corn to grade it as with the conventional corn and could therefore pack much faster.

Segregation was difficult with the sweet corn right from picking to storage. Other difficulties observed were the need to keep detailed and accurate records on where each variety was planted, and to have separate containers for holding and transporting the corn from the field. Farm workers also had to be trained and monitored to ensure that they kept track of which variety they were picking and labelled each accordingly.

## Marketing of Corn

The sweet corn was available for purchase in the market at Birkbank Farms in Orton and fully labelled along with information on the number of sprays used and relative costs to produce. The two types of corn were presented in separate wooden bins marked either "genetically engineered Bt sweet corn" or "regular sweet corn." The non-Bt corn was labelled "regular" as opposed to "non-Bt" or "conventional" for simplicity. There was approximately 30 centimetres of space between the two bins The space was filled with bags and pamphlets to prevent the corn from becoming inadvertently mixed. Employees in the market kept both corn bins filled to the same level throughout the day. The genetically engineered Bt corn and the conventional corn were both sold for the same price, $3.99/dozen. Pamphlets with background information were also available in the market for customers to take home.

On weekends, free samples of both conventional and Bt sweet corn were distributed in front of the market and a short questionnaire was conducted as an intercept interview. It consisted of four multiple-choice type questions, one open-ended question, and five demographic questions. The questionnaire was pilot tested at the farm on several customers and it was determined that five questions were ideal in order to avoid overburdening the participants. The main questions assessed current awareness of genetically engineered foods; whom they trusted to provide them with information; what influenced their choice to buy Bt or non-Bt sweet corn; and whether they were more concerned with pesticides or genetic engineering. The inclusion criteria for participants were that they had to buy corn in the market or be offered the choice to consume Bt or non-Bt corn. Once they had made their choice and purchased the corn they were approached and asked to answer the questionnaire.

Observations were also taken while the customers were making their decisions. Any questions the respondents asked about genetically engineered foods in the market were recorded, as well as their comments. Basic content analysis was conducted on the comments to obtain a sense of customers' initial reactions to the market display and the option of purchasing genetically engineered foods. Comments were coded as either positive, negative, neutral or questioning, then each category was coded for six main themes. These were: pesticides, environment, taste/quality, apathy, testing, emotional response, and lack of information. The themes were selected after going through the data and identifying descriptive categories that were representative of the data. The coding system and data were verified by another researcher. During sampling, many people who tried the samples also had questions about the technologies and these questions were recorded.

## Results of Study

Sales of both types of corn were recorded from 30 August 2000, when the corn was first harvested to 6 October 2000, at which time the conventional sweet corn could no longer be sold due to a shorter shelf-life. Small amounts of the Bt sweet corn were sold over the next two weeks; however, these were not included in the sales because although they were labelled, there was no conventional corn available for comparison. Bt sweet corn sales were consistently higher than regular sweet corn sales. In total, over the five-week period, 136 bags or 8,160 cobs of Bt corn were sold and 90.5 bags or 5,430 cobs of conventional corn.

A total of 174 consumers were surveyed at the three venues. Of these, 89 percent had at least heard or read about genetically engineered foods. Having heard about GE foods, however, did not have an influence on their perception of whether they were harmful or beneficial. Forty-nine percent of consumers surveyed said GE foods were beneficial, 11 percent said they were harmful, 5 percent said they were both beneficial and harmful, and 36 percent said they did not have enough information to form an opinion.

When asked whether they would buy GE corn (or had bought GE corn in the case of Birkbank Farm customers), 69 percent of all surveyed said they would buy it, 26 percent said they would not, and 5 percent did not know. Of those who had heard or read about GE foods, 68 percent said they would buy it. When asked why they would buy GE food, the top answers were taste/quality and fewer pesticides. Among those who would not buy GE food, the primary reasons were environmental concerns, health and safety, ethical reasons, and a perceived need for additional testing.

When asked whether they were more concerned about pesticides or genetic engineering, the majority (72 percent) felt that pesticides posed more of a threat, 13 percent were equally concerned about both, 5 percent were more concerned about genetic engineering, and only 3 percent were not concerned about either. When examining the effect of attitude on purchasing decisions, 87 percent of those who said that GE was beneficial would buy genetically engineered foods, while 74 percent of those who said it was harmful would not purchase GE foods.

A total of 115 comments were recorded over 13 weekend days while free samples of both types of corn were being given away. The majority, 30 percent of all comments recorded in the market, were positive toward Bt sweet corn. Questions made up the next largest component at 24 percent and there were an equal number of negative and neutral comments (22 percent in each case). Fifty percent of all comments were related to taste and quality. This is not surprising as the comments were often recorded while the respondent was tasting the corn.

Of the positive comments, the majority was about taste and quality. The second most common topic was pesticide use. For example, a common statement was "I like the fact that they used fewer insecticides on the Bt one." The remaining comments that did not fall into one of the two categories above were classed as apathy — that is they were all along the lines of "I don't really care if its genetically engineered or not."

With regard to the negative comments, most of these were, again, taste/quality comments such as "I think the regular corn tastes better." The rest of the comments were split between comments about environmental concerns and emotional responses. Comments relating to environment were primarily about gene flow: for example, "I am worried about the genes escaping into the wild." Emotional responses were comments where the respondent could not articulate a reason for his or her statements: for example, "I am nervous about genetically engineered foods, I don't even want to try it."

## Discussion

Taste and quality appear to be important attributes for consumers when purchasing fresh market sweet corn, especially when price was not a factor. The fact that the majority of the comments were positive is also consistent with the sales figures.

Both the survey data and qualitative comments indicate that the main influences over the customers' decision to buy Bt corn were taste and quality, decreased pesticide use, and curiosity. Many people opted to buy both or just purchased a small number of the Bt corn to "try it out." Research has indicated that consumer acceptance of GE foods will be based on the recognition of tangible benefits (Bruhn *et al.* 1992; Decima Research 1993; Frewer, Shepherd and Sparks 1994; Hoban 1997; Sheehy 1996; Zimmerman *et al.* 1994). The data presented here indicate that some of the customers at the Birkbank Farm market felt that a reduction in pesticide use was a tangible benefit that may have influenced their decision to purchase the Bt sweet corn. Canadians have been shown to be very concerned about the use of chemical pesticides in agriculture, having rated pesticides a more serious hazard to food safety than genetic engineering (CFIA 1998).

The information boards associated with the produce may have influenced consumers to buy just because of the board's presence or perhaps because there was detailed information provided. A few customers in the market were observed to fill their bags with the regular corn and then pause to read the large signs above the bins, which explained the pest management regime for each type of corn. They then proceeded to empty their bags and refill them with the Bt sweet corn. This example illustrates that pesticide reduction is something that some consumers at Birkbank Farms value and that labelling foods as GE and non-GE, or Bt and non-Bt may not aid consumer decision-making. As previously described, many customers at the Birkbank Farm market knew very little about GE foods, especially when it came to specific benefits and risks. At the same time it seemed that many customers had little knowledge of how conventional crops were produced and the associated pesticide use.

The primary concerns, identified through surveys, were mostly related to environmental considerations, specifically gene flow. Other particular concerns mentioned included unknown future consequences and that the products were not adequately tested. Research on public perceptions has identified human and environmental health as the top concerns over GE foods among Canadians. It also reflects the major media coverage on Bt sweet corn, which has focused primarily on environmental effects of GE crops, especially effects on non-targets such as the Monarch butterfly.

Because this was a commercial farm, much of the produce grown was sold to larger retail outlets. As the project was completely open, retailers were well aware that genetically engineered Bt sweet corn and Bt potatoes were grown at Birkbank Farms. Noticeably, retail produce buyers did not ask whether the corn and potatoes they were purchasing from the farm were GE or not, nor did they want to know. When approached to see if they would allow a product information display similar to the one at the Birkbank Farm market for consumer research in a retail setting, retailers were not willing to draw such attention to themselves. This is not surprising because special interest groups had been targeting large retail chains in urban centres as part of their demonstrations and protests. In this environment, retailers were more hesitant to draw extra attention to their operation. With the extensive media coverage at the farm, some customers at retail chains were asking the store managers whether the corn and potatoes were genetically engineered. Only then would some of the retailers call the farm and ask.

This research is a starting point and describes the experience on one farm during one year. In order to develop a truer sense of both grower and consumer acceptance of genetically engineered crops, this research should be continued over several years and involve other growers and retail outlets. As the weather in southern Ontario was sufficiently poor during this experiment that commercial organic sweet corn was not available, a comparison with organic growing methods and organic produce would also be useful in terms of comparing the costs and benefits of the different methods and consumer preference.

## CONCLUSION

Slovic noted, "we live in a world in which information, acting in concert with the vagaries of human perception and cognition, has reduced our vulnerability to pandemics of disease at the cost of increasing our vulnerability to social and economic catastrophes of unprecedented scale. The challenge before us is to learn how to manage stigma and reduce the vulnerability of important products, industries, and institutions to its effects, without suppressing the proper communication of risk information to the public" (1997).

Producer-led risk management programs are an appropriate risk-management strategy to demonstrate to consumers that producers are cognizant of consumer concerns about food safety and to demonstrate that producers and others in the farm-to-fork continuum are working to reduce levels of risk. Appropriate risk management strategies, such as on-farm food safety programs, as well as an honest discussion with consumers about agricultural technologies using a demonstration farm and market, are tools to enhance consumer confidence while producing scientifically safe food.

## NOTES

1.   The Web site is <www.plant.uoguelph.ca/safefood>.

## REFERENCES

Angus Reid Group Inc. 1995. "Public Opinion on Food Safety and Biotechnology Applications in Agriculture," *Angus Reid poll.*

Bruhn, C., S. Peterson, P. Phillips and N. Sakovidh. 1992. "Consumer Response to Information on Integrated Pest Management," *Journal of Food Safety* 12:315-26.

Canadian Food Inspection Agency (CFIA). 1998. *Safe Food Handling Survey.* Ottawa: CFIA.

Decima Research. 1993. *Final Report to the Canadian Institute of Biotechnology on Public Attitudes Towards Biotechnology.* Toronto: Decima.

Einsiedel, E.F. 1997. *Biotechnology and the Canadian Public.* Report on a 1997 National Survey and Some International Comparisons. Calgary: University of Calgary.

_____ 2000. "Cloning and its Discontents: A Canadian Perspective," *Nature Biotechnology* 18:943.

Frewer, L.J., R. Shepherd and P. Sparks. 1994. "Biotechnology and Food Production: Knowledge and Perceived Risk," *British Food Journal* 96(9):26-32.

Gregory, R., P. Slovic and J. Flynn. 1995. "Risk Perceptions, Stigma, and Health Policy," *Health and Place* 2(4):213-20.

Hagedorn, C. and S. Allender-Hagedorn. 1997. "Issues in Agriculture and Environmental Biotechnology: Identifying Biotechnology Issues from Public Opinion Surveys, the Popular Press and Technical/Regulatory Sources," *Public Understanding of Science* 6:233-45.

Hedberg, C.W., K.L. MacDonald and M.T. Osterholm. 1994. "Changing Epidemiology of Food Borne Disease: A Minnesota Perspective," *Clinical Infectious Diseases* 18:671-82.

Hoban, T.J. 1997. "Consumer Acceptance of Biotechnology: An International Perspective," *Nature Biotechnology* 15:232-34.

Peterson, R. 2000. "Public Perceptions of Agricultural Biotechnology," and "Pesticides: Recent Understandings and Implications for Risk Communication," *American Entomologist* 46:8-16.

Phillips, P.W.B. and G. Isaac. 1998. "GMO Labeling: Threat or Opportunity?" *AgBioForum* 1(1).

Powell, D.A. 2000. "Food Safety and the Consumer: Perils of Poor Risk Communication," *Canadian Journal of Animal Science* 80(3):373-404.

Rushing, J.W., F.J. Angulo and L.R. Beuchat. 1996. "Implementation of a HACCP Program in a Commercial Fresh-Market Tomato Packinghouse: A Model for the Industry," *Dairy, Food and Environmental Sanitation* 9:549-53.

Sheehy, H. 1996. *Consumers and Biotechnology: A Synopsis of Survey and Focus Group Research.* Ottawa: Office of Consumer Affairs.

Slovic, P. 1997. "Perceived Risk, Stigma, and the Vulnerable Society." Paper presented at Conference on Risk, 13 June, City University, London.

Tauxe, R.V. 1997. "Emerging Foodborne Diseases: An Evolving Public Health Challenge," *Emerging Infectious Diseases* 3(4). <http://www.cdc.gov/ncidod/eid/vol3no4/tauxe.htm>.

United States. National Academy of Sciences' National Research Council Committee on Risk Characterization. 1996. *Understanding Risk: Informing Decisions in a Democratic Society*. Washington, DC: National Academy Press.

United States. Presidential/Congressional Commission on Risk Assessment and Risk Management. 1997. "Framework for Environmental Health Risk Management," *Final Report*, Vol. 1. Washington, DC. <http://www.riskworld.com>.

Zhuang, R.Y., L.R. Beuchat and F.J. Angulo. 1995. "Fate of *Salmonella montevideo* on and in Raw Tomatoes as Affected by Temperature and Treatment with Chlorine," *Applied and Environmental Microbiology* 6:2127-31.

Zimmerman, L., P. Kendall, M. Stone and T. Hoban. 1994. "Consumer Knowledge and Concern about Biotechnology and Food Safety," *Food Technology* 71-77.

# 11

# Genetic Seeds of Discord: The Transatlantic GMO Trade Conflict after the Cartagena Protocol on Biosafety

*Robert Falkner*

## INTRODUCTION

Food policy and biological diversity are interlinked in a complex relationship. On the one hand, biodiversity is an important factor in sustaining a diverse system of food production. Preserving the genetic riches of ecological systems is for many countries an integral part of their food security strategy. In this sense, biodiversity protection and food security are interdependent and mutually supportive. On the other hand, modern food production can negatively impact on the natural environment. The industrialization of agriculture has brought with it serious ecological hazards, such as the increased use of herbicides and pesticides. More recently, the introduction of biotechnological developments to agriculture has been seen as a potential threat to biodiversity. In other words, modern food security strategies can endanger biological diversity.

Recognizing these complex linkages between food policy and biodiversity, the international community has addressed some of the resulting policy challenges in a number of forums. In the area of agricultural biotechnology, for example, the Food and Agriculture Organization (FAO) and the United Nations Environment Program (UNEP) have created codes of conduct on safety in biotechnology, and the parties to the Convention on Biological Diversity (CBD) initiated a negotiation process on a binding international treaty that deals with safety aspects of international trade in genetically modified organisms (GMOs).

From a food security perspective, efforts to protect biodiversity have a growing impact on the production and international trade in food. The CBD's protocol on the safety of biotechnology is a prime example of how international environmental policy-making is affecting food and trade policy across the world. The Cartagena Protocol on Biosafety, negotiated in the second half of the 1990s and adopted in January 2000, establishes a system of information-sharing and risk assessment that allows countries to prohibit the import of living modified organisms (LMOs)[1] if they are found to pose a risk to biodiversity or human health. The Protocol thus legitimizes interference with international agricultural trade in the name of environmental or health protection. While welcomed by environmentalists, the agreement has caused concern among agricultural exporters that it might create new barriers to trade in agriculture (Papanikolaw 2000).

This chapter focuses on the relationship between the emerging international biosafety regime and international agricultural trade. It examines the controversy surrounding agri-biotechnology and assesses the potential impact of the Cartagena Protocol on the transatlantic trade conflict over genetically modified (GM) crops and food. The subsequent discussion is divided into three parts.

First, the transatlantic dispute over agricultural biotechnology will be examined, with a particular focus on why it is that policymakers in the United States, Canada, and the European Union (EU) have found it so difficult to bridge their differences in regulating GM food. It is argued that the GM trade conflict is not a traditional trade dispute. Rather, it concerns trade barriers resulting from a divergence in domestic regulatory priorities and practices. While such non-tariff trade barriers have troubled advocates of trade liberalization for some time, they have become more relevant in recent years. What is at stake in these conflicts is not merely competing commercial interests but also, and arguably more importantly, differences in societal values and preferences. The GM trade conflict is not simply about the correct interpretation and application of World Trade Organization (WTO) rules. It is also about the legitimacy of the WTO to arbitrate in cases where societal and cultural differences shape trade policy.

Second, the creation of the BioSafety Protocol and its major elements will be analyzed. Against the background of deteriorating transatlantic trade relations, the fact that an agreement on the Protocol was reached represents in itself a significant political achievement. It has gone some way in reassuring those who worry about the erosion of national regulatory power in an era of seemingly unfettered globalization. The Protocol includes several innovative features, including explicit references to precaution in public policy decision-making.

Finally, the question of what likely impact the BioSafety Protocol will have on the transatlantic relationship will be examined. It was hoped that the international accord would create a more stable and harmonized regulatory environment for agricultural biotechnology. But as will be argued below, it is highly questionable whether the provisions of the biosafety treaty will indeed help to resolve the conflict. More likely than not, they will have no major impact.

## THE TRANSATLANTIC GMO TRADE CONFLICT

Genetically modified crops are being adopted by a growing number of countries, most notably the United States and Canada, but also Argentina and China. The first commercial plantations of GM crops occurred in the mid-1990s. Since then, agri-biotechnological production has grown at a rate unrivalled by any other technological innovation in crop development. The global area of GM crop production has risen from 1.7 million hectares in 1996 to more than 44 million hectares in 2000 (see Chapter by Phillips for details). Given strong growth performance in the last five years and the ongoing R&D efforts in agricultural biotechnology, market analysts reckon that the global market for GM crops may reach $8 billion in 2005 and $25 billion in 2010 (James 1999).

Most of the commercially relevant R&D in this sector has been carried out by a small number of multinational firms such as Monsanto and Novartis, who have developed GM crops with a view to commercializing them worldwide. But while North American regulators have given a green light to many new GM crop varieties early on, the European authorities have taken a more cautious approach. As a consequence, the US and Canada have come to dominate worldwide production of GM crops, while less than 1 percent of the world's output of GM crops originates from EU countries.

At the centre of the transatlantic GM conflict is the EU's *de facto* moratorium on the commercial release of GM crops into the environment and ban on imports of GMOs and GM products. North American trade officials claim that this constitutes a breach of international trade obligations, while the European Union insists that the long-term risks posed by GMOs are uncertain, and that therefore precaution should guide policy-making in this area.

The EU's regulatory system for GMOs consists of four main pieces of legislation, covering the contained use of GMOs (Directive 90/219); the release of GMOs into the environment and their placement on the market (Directive 90/220); the approval of Novel Foods (Regulation 258/97); and mandatory labelling of all GM products not covered by the preceding legal instruments (Regulation 1139/98). The approval procedure for the bringing to market of GMOs has been criticized for being cumbersome and open to abuse by politically motivated member states who wish to prevent GMO approvals without a sufficient scientific basis to support their case. Likewise, the labelling requirements for GM products stand accused of unnecessarily branding a technology as potentially dangerous when in fact specific products and their potential risks should be the primary concern of regulators. From a North American perspective, where biotechnological processes are not considered to be substantially different to other plant-breeding methods, the European focus on process-based labelling appears as fundamentally flawed and prone to giving rise to trade barriers.[2]

With growing public concern over food safety and the introduction of GM crops in the second half of the 1990s, the EU's regulatory approval system for GMOs has virtually ground to a halt. When in 1996–97 the first shipments of GM maize and soybeans arrived in Europe from North America, environmental and consumer campaign groups began targeting the growing trade in GMOs and set off an intense public debate in most EU member states, but particularly in the United Kingdom. At the same time, regulatory authorities in the EU began to reconsider their position on approvals for the commercial use of GM crop varieties. Several GMOs, including Bt-maize, had been approved under Directive 90/ 220 until 1997. However, the EU imposed a *de facto* moratorium on new authorizations in 1998 which continues to remain in place.

Given that the EU has found it hard to reform its regulatory regime — which the European Commission itself admits is unsatisfactory — it is not surprising that the United States and Canada should continue to accuse the EU of practising protectionism. Faced with growing domestic pressure over the loss of crop export markets in Europe, they insist that the problem lies entirely with the EU's regulatory system. Given other long-standing disputes over EU bans and restrictions on agricultural imports (e.g., hormone-treated beef, bananas), the conflict over GM crops is as much about the commercial interests of affected farmers as about the principal of resisting growing agricultural protectionism worldwide.

Yet, despite the apparent similarity of the GM dispute with conventional trade conflicts, the charge of protectionism does not go to the heart of the problem. The EU's GM policy may result in measures that appear to be of a protectionist nature. But it is important to recognize that the policy is motivated by consumer concerns, not protectionist interests. In other words, the EU's regulatory system has protectionist *effects* — it harms North American farming export interests — but EU policy is not driven by protectionist *interests*. European GM policy has evolved in response to growing public fears about food safety and the environmental risks involved in agricultural biotechnology.

It is this societal dimension that distinguishes the GM trade conflict from other areas of contention, such as EU subsidies to Airbus production or the European bananas import regime. A significant part of European society feels that the free trade principle should not be allowed to overrule the sovereign right of democratic states to determine what levels of environmental and health risks they wish to tolerate. Fears about GM crops and foods may be based on an incomplete understanding of the scientific aspect of biotechnology. The crucial point here is that in the absence of consensual scientific opinion, societies need to weigh the benefits and risks involved in adopting new technologies, and that cultural or political differences between societies will influence these decisions. This is not the place to discuss the environmental safety of GM crops, but it is worth noting that the scientific community is far from united in its views on this subject.

It is these cultural and political differences that make it difficult, if not impossible, to harmonize rules at the international level regarding trade in agricultural biotechnology. Trade diplomacy, and the WTO as the final arbiter in trade disputes, is often at a loss when it comes to negotiating between different societal value systems that inform trade policy. The General Agreement on Tariffs and Trade (GATT)/WTO system has found it relatively easy to overcome protectionist producer interests throughout its 50-year history, but has failed to come up with solutions for dealing with conflicting interests that reflect incompatible societal preferences.

North American policymakers have so far failed to come to terms with this aspect of the GMO trade conflict, but seem to have accepted that the WTO is not the right instrument for resolving this conflict. In the past, US trade officials in particular have repeatedly threatened to invoke WTO obligations if the EU does not drop its *de facto* ban on GM crops or change its GM labelling regime (Brown 2000). But these threats were often made more with a view to placating domestic interest groups and a weary Congress than with a clear intention of bringing a GMO case to the WTO.

Overall, the Clinton administration applied a gradualist approach to this problem, keeping up the pressure on the EU while sponsoring efforts to promote a shift in the public debate from the risk of biotechnology to its benefits. In the run-up to the 1999 WTO Ministerial Conference in Seattle, the United States, together with Canada and Japan, proposed the creation of a WTO working group on biotechnology, in the hope of shifting the debate to consider and clarify the trade obligations in this area. But EU resistance against this proposal prevented the WTO from holding full-scale discussions on biotechnology. And the US administration, fully aware of the potentially explosive consequences of having the WTO decide a GM conflict, has so far abstained from making a formal complaint to the WTO about the EU's regulatory system.[3]

The collapse of the WTO meeting in Seattle further underlined the sensitive nature of efforts to introduce new environmental concerns to the WTO system. After the US and Canada failed to open WTO talks on biotechnology, the focus in the transatlantic GM conflict quickly shifted to a different international forum: the final round of negotiations on the BioSafety Protocol, which was held in Montreal in January 2000, two months after the Seattle conference.

## THE CARTAGENA PROTOCOL ON BIOSAFETY

International efforts to regulate biosafety go back to the 1980s. In the preparatory meetings for the 1992 United Nations Conference on Environment and Development (UNCED) biotechnology and biosafety were introduced by developing countries as areas of concern to be covered by the Convention on Biological Diversity. After protracted negotiations, in which the United States signalled that it could not sign the final compromise, the CBD was adopted in 1992 with a provision that called upon the parties to consider the need for and modalities of a protocol on biosafety. It took another three years until the international community agreed on a negotiating mandate, and in 1996 the open-ended ad hoc Biosafety Working Group (BSWG) began its work of drafting an international biosafety treaty.[4]

The Biosafety Working Group met six times between 1996 and 1999, producing a list of provisions for inclusion in the Protocol. The BSWG meetings failed, however, to come up with a draft treaty text that was acceptable to all countries. When the Conference of the Parties to the CBD met for an extraordinary session in Cartagena (Colombia) in February 1999 (the so-called ExCOP), delegates were faced with the almost impossible task of eliminating a vast amount of so-called "bracketed text" in the draft treaty that contained outstanding disagreements between the parties. In the end, the Cartagena conference collapsed and the parties decided to reconvene the ExCOP within 15 months to try to bring the negotiations to a successful conclusion.

The conflict at Cartagena centred on several elements of the biosafety regime. There was, first of all, the question of the scope of the treaty, which pitted industrialized countries against developing countries. The south demanded an all-inclusive scope that included all types of living modified organisms and LMO usage. Fearful of the new technology and lacking in domestic regulatory capacity, many developing countries expected the BioSafety Protocol to provide them with the legitimacy and means to control all forms of LMO trade. In contrast, northern countries felt that such a wide scope would create unnecessary regulatory burdens in areas that posed no threat to the environment. This was seen to be the case with LMOs as pharmaceuticals, LMOs destined for contained use, and LMOs in transit. The United States and Canada in particular were concerned that a wide scope of the protocol would only serve to hinder international trade in biotechnology products. They argued that the Protocol should focus on the transboundary movement of LMOs that were destined for direct release into the environment, as is the case in the international seeds trade.

A second area of contention was the inclusion of agricultural commodities in the Protocol's regulatory mechanism, the advance informed agreement (AIA). The developing countries, this time with the support of the European Union, wanted to have trade in genetically modified crops treated in the same way as other LMOs covered by the treaty. This issue had only become virulent half-way through the biosafety talks, after countries like the United States and Canada had started to introduce GM crops on a commercial basis and the European Union began to block North American GM soybean and corn shipments. By the time of the Cartagena conference, the question of LMO commodities had moved centre-stage in the negotiations. Given the commercial stakes involved for the group of GM crop-exporting countries, the commodities issue became a major stumbling block on the way to a successful conclusion of the biosafety talks.

Other contentious aspects of the Protocol included demands by developing countries for liability and redress provisions and for the right to block LMO imports for socio-economic reasons, the insistence by the European Union and the south that importing nations should be able to take precautionary action when carrying out risk assessment, and European demands for an identification scheme for GM commodities. In all of these areas of conflict, the group of GMO-exporting countries (the Miami Group, consisting of the US, Canada, Argentina, Australia, Chile, and Uruguay) was on one side of the debate, arguing for minimal and least trade-restrictive provisions with either the European Union or the group of developing countries (the Like-Minded Group), and in some cases both of them pushing for a stricter and more comprehensive regulatory regime.

By the time of the Cartagena conference, the rise of anti-GM food campaigns in Europe had transformed the international biosafety negotiations. What had originally begun as a north-south conflict — with developing countries pushing for a comprehensive biosafety regime and the northern countries seeking to protect their biotechnology industry from burdensome regulation — soon developed into a transatlantic conflict. Against the background of eroding public confidence in European food safety regulation in the wake of mad cow disease, the EU imposed a moratorium on new approvals for GMOs and began drafting legislation on the labelling of GM food products. At the same time, European negotiators were arguing for a stricter biosafety treaty that would allow nations to make decisions on GMO imports on the basis of the precautionary principle. Moreover, the Europeans insisted that the BioSafety Protocol should not be subordinate to the disciplines of the WTO.

The Cartagena conference saw an increasing polarization of positions, with the EU and the Miami Group locked into an increasingly bitter conflict over the nature and scope of the biosafety regime. The US and Canada felt that the European position was driven more by a concern for appeasing anti-GMO sentiment at home than an interest in creating a workable regulatory system. They feared that the EU was using the biosafety negotiations to pursue wider political objectives, namely to shift the balance between trade rules and environmental agreements in favour of the latter. Precautionary action as demanded by the Europeans was interpreted by the Miami Group as allowing importing nations to extricate themselves from the science-based risk assessment as demanded by WTO rules. This, industry and agricultural lobbyists insisted, had to be prevented.

The transatlantic conflict came to a head in Cartagena, when the Miami Group rejected an eleventh hour compromise package tabled by the EU and supported by the other negotiating groups. The collapse of the conference caused world-wide interest in the biosafety talks and propelled biosafety issues into a more prominent position on the international agenda. Furthermore, the year 1999 saw an unprecedented backlash against GMOs in Europe, and even in North America environmental non-governmental organizations (NGOs) began paying more attention to biosafety in their campaigns.

Despite the mutual recriminations of the parties, informal meetings continued after Cartagena to rebuild the necessary trust among negotiators for reaching a final agreement. Although they failed to resolve any of the key issues that had led to a breakdown of the talks, the informal meetings helped to put the international process back on track. In another context, efforts by the United States and Canada to establish a WTO working group on biotechnology failed at the end of 1999, and together with the collapse of the Seattle WTO ministerial meeting in December, this only contributed to the growing momentum for a compromise in the biosafety talks.

At the resumed ExCOP in Montreal in January 2000, the parties were able to reach an agreement only after intense day-and-night negotiating sessions which, in their final phase, also involved several environment ministers. The transatlantic differences with regard to the role of precaution and the relationship between the biosafety treaty and other international agreements were among the last stumbling blocs in this process. In many ways, the BioSafety Protocol represents an incomplete compromise between the Miami Group and the EU, as well as the developing countries. Several instances of ambiguous treaty language allowed each group to claim victory, and several key aspects are to be resolved in the near future, after the agreement has entered into force.

It is impossible here to give a complete overview of all the treaty's provisions (See Bail, Falkner and Marquard 2002). Suffice it to say that the principal objective of the treaty is to accord the right to nations to decide on whether or not to accept GMO imports on the basis of the advance informed agreement. The AIA mechanism obliges exporters of GMOs to provide advance information on their products so that importers can make an informed decision based on scientific risk assessment. In making such decisions, importing nations may err on the side of caution:

Lack of scientific certainty due to insufficient relevant scientific information and knowledge regarding the extent of the potential adverse effects of a living modified organism on the conservation and sustainable use of biological diversity in the Party of import, taking also into account risks to human health, shall not prevent that Party from taking a decision, as appropriate, with regard to the import of the living modified organisms in question... (article 10(6), Cartagena Protocol on Biosafety).

With regard to the relationship of the BioSafety Protocol to the WTO and other international agreements, the treaty contains a compromise formula in the preamble which states the mutual supportiveness of trade and environment agreements while protecting rights and obligations under existing international agreements. A second area of contention was the inclusion of agricultural commodities in the protocol's regulatory mechanism, the AIA on the first transboundary movements of LMOs intended for deliberate release. While including commodities in the scope of the protocol but excluding them from the AIA procedure, the parties made them subject to a comparatively weaker mechanism that mandates information-sharing on the commercial introduction of commodities (article 11). On the insistence of the EU, exporters of LMO commodities are required to identify them in their accompanying documentation as "may contain" LMOs (article 18(2)a), a provision that is to be specified in future negotiations (article 18(3)).

## THE IMPACT OF THE BIOSAFETY TREATY ON THE TRANSATLANTIC GMO TRADE CONFLICT

What impact on the transatlantic GMO trade conflict is the Biosafety Protocol likely to have? While it is too early to answer this question with certainty — the Protocol will only come into effect once 50 nations have ratified it, and it will take many more years thereafter to negotiate some of its unresolved elements such as labelling and liability — it is possible to outline possible scenarios for the future.

First, reaching an agreement in Montreal has contributed to calming tempers in the transatlantic trade conflict. Both sides have welcomed the conclusion of the biosafety talks and have declared their willingness to cooperate in the implementation of the agreement. The United States, being a non-party to the CBD, will not be required to introduce domestic implementation legislation, and there are currently no signs that the US Congress will ratify the Convention on Biological Diversity, a precondition for signing and ratifying the BioSafety Protocol. However, US policymakers have cooperated with the interim process of the Intergovernmental Committee for the Cartagena Protocol (ICCP) and other initiatives such as in the area of capacity-building, which will pave the way for implementation once the treaty enters into force.

Beyond having restored mutual understanding and good-will among the parties, however, the biosafety agreement has not significantly altered the main protagonists' perceptions of the key areas of contention. Reading interpretative statements issued by Miami Group members and the European Union after the adoption of the Biosafety Protocol, one is led to conclude that the treaty represents little more than a truce in a battle that will continue to sour transatlantic trade relations.

After the conclusion of the resumed ExCOP in Montreal, the EU was quick to claim that its position on GMO regulation had been strengthened by the biosafety agreement. The Commission argued that the treaty legitimized its own regulatory system by incorporating the precautionary principle in its operational part and that the compromise on the relationship of the biosafety treaty with the WTO further reduced the chances of a challenge to its GMO regulations brought under the WTO system. Both these claims are rejected by Canadian and US negotiators and policymakers, who argue that the treaty does not include the precautionary principle as interpreted by Europeans, nor does it substantively change the rights and obligations under the WTO.

With regard to the role of precaution, the central disagreement concerns the interpretation of article 10(6) of the BioSafety Protocol, which states that "lack of scientific certainty due to insufficient relevant scientific information and knowledge" should not stand in the way of an importing nation's decision. Canada and the United States insist that this formulation does not establish the precautionary principle as such, but merely states common practice of regulators in circumstances of uncertain scientific evidence. In a letter to US farming representatives, Charlene Barshefsky, US trade representative in the Clinton administration, stated that this provision "does not undermine WTO disciplines. The language in the final text of the Protocol essentially states a truism" (Barshefsky 2000).

Given this interpretation, it comes as no surprise that US and Canadian officials continue to accuse the EU of abusing the pecautionary principle for protectionist purposes. Thus, the US and Canada are not retracting from a common position on the role of precaution reached in the biosafety talks, but simply deny that the European interpretation ever formed the basis for the compromise deal. They view the BioSafety Protocol's reference to precaution as being consistent with WTO obligations.

In a similar fashion, the North American position on the relationship of the Protocol with international agreements, and in particular the WTO, remains unchanged. The fact that the final treaty text does not contain a "savings clause," which would guarantee that WTO rules apply in a GMO-related trade dispute, is not seen as undermining the interests of the GMO-exporting nations. Again, as in the case of the text on precaution, the treaty's ambiguous wording on the "relationship" issue gives rise to conflicting interpretations. The compromise text in the treaty's preamble states:

*Recognizing* that trade and environment agreements should be mutually supportive with a view to achieving sustainable development;

*Emphasizing* that this Protocol shall not be interpreted as implying a change in the rights and obligations of a Party under any existing international agreements;

*Understanding* that the above recital is not intended to subordinate this Protocol to other international agreements. (BioSafety Protocol, Preamble)

Adopting this compromise formula, which was borrowed from the Rotterdam Convention on the Prior Informed Consent Procedure for Certain Hazardous Chemicals and Pesticides in International Trade, opened the way to reaching a compromise in the final negotiations. While it may have been a necessary diplomatic fudge to move the negotiation process forward, the preambular text fails to clarify the precise relationship between trade and biosafety rules. The US and Canada insist that the BioSafety Protocol does not close off the possibility of bringing a case against the EU's GMO regulations at the WTO, while European Union officials see the biosafety treaty as an important step toward shifting the balance between trade rules and environmental treaties in favour of the latter.

Lawyers on both sides will continue to argue over the correct interpretation of the "trade relationship" provision in the biosafety treaty,[5] but it seems clear that a potential escalation of the transatlantic GMO trade conflict involving a formal complaint within the WTO against the EU cannot be ruled out. US officials have in the past repeatedly threatened to take such action, especially against the EU's GMO labelling regime which they regard as erecting illegitimate non-tariff trade barriers. They will want to keep this option open for future efforts to bring about a change in the EU's GMO regulations.

However, despite the often belligerent rhetoric, both the US and Canada have so far appeared to be keen to avoid an open confrontation with the EU involving the WTO's dispute settlement mechanism, well aware of the potentially explosive implications this might have. It is widely recognized that, given the strong political sensitivities involved, the WTO's legitimacy in the public's eye could easily be undermined if it had to decide a GMO-related case. Both North American interest groups and policymakers seem to acknowledge this and appear not to want to destabilize public trust in the WTO system when they are still hoping to address larger and commercially more important issues in the next WTO trade round, such as reducing EU subsidies to agriculture.

The question of whether the United States or Canada is likely to bring a case against the EU's GMO regulation may therefore be more a question of political expediency than of legal interpretation. The BioSafety Protocol has failed to resolve the central issues in the transatlantic GMO conflict, and arguably was never in a position to do so. It may have nudged the "balance of power" in the international debate on the role of precaution and the relationship between the WTO and environmental agreements slightly in the EU's preferred direction, though without significantly undermining the GMO-exporting countries' position.

## CONCLUSION

The regulation of agricultural biotechnology has become a highly sensitive issue in the transatlantic trade relationship. European concerns over the environmental and human health risks of genetically modified crops have contributed to an increasingly restrictive regulatory regime in the EU, which in turn has led to North American complaints about rising EU protectionism and a breach of WTO obligations by EU regulators. The successful conclusion of the BioSafety Protocol negotiations has done little to resolve the central areas of contention in the transatlantic GMO conflict. Far from it, American, Canadian, and European policymakers continue to disagree on the role of precaution in biosafety regulation as well as the relationship between the BioSafety Protocol and the WTO's trade rules. The stakes in this conflict are not simply to do with competing commercial interests; differing societal preferences and values shape trade policy, thereby making it more difficult for trade diplomats to reach an international compromise. In that the GMO trade conflict does not represent a traditional trade dispute centred on producer protectionism, it is likely to prove very difficult to resolve within the WTO's framework for dispute settlement.

### NOTES

The author thanks government officials and BioSafety Protocol negotiators from Canada, the United States, and the European Union for their willingness to be interviewed. Their requests for anonymity are respected.

1. The Cartagena Protocol uses the term living modified organism (LMO) instead of the more frequently used genetically modified organism (GMO) or organisms derived from genetic engineering.
2. On the EU's GM regulations, see The Open University (2000).
3. There has been no discernible change of policy under the new Bush administration, but it is too early to judge its intentions in the biosafety area.
4. For a brief overview and analysis of the biosafety negotiations, see Falkner (2000). A more in-depth discussion of the Cartagena Protocol can be found in Bail, Falkner and Marquard (2002).
5. See, for example, the differing views expressed by Margarido Afonso and Sabrina Safrin in Bail, Falkner and Marquard (2002).

REFERENCES

Bail, C., R. Falkner and H. Marquard, eds. 2002. *The Cartagena Protocol on Biosafety: Reconciling Trade in Biotechnology with Environment and Development?* (London: RIIA/Earthscan), forthcoming.

Barshefsky, C. (US Trade Representative). 2000. Letter to Bob Stallman, president of the American Farm Bureau Federation, 6 March.

Brown, P. 2000. "New Trade War Looms over GM Labelling," *The Guardian* (London), 31 July.

Falkner, R. 2000. "Regulating Biotech Trade: The Cartagena Protocol on Biosafety," *International Affairs* 76(2):299-313.

James, C. 1999. *Global Status of Commercialized Transgenic Crops, 1999*, ISAAA Briefs No. 12: Preview (Ithaca, NY: ISAAA).

The Open University. 2000. "EU Safety Regulation of Genetically-Modified Crops." Summary of a ten-country study funded by DGXII under its Biotechnology Program. <www-tec.open.ac.uk/bpg.htm>.

Papanikow, J. 2000. "Biosafety Protocol Receives Mixed Reception from Agricultural Groups," *Chemical Market Reporter* (New York), 7 February.

# Conclusion

# 12

# Governing Food: Closing Remarks

*William Leiss*

## INTRODUCTION

The conference brochure provided the following setting for the discussions:

> Food safety is a dramatic example of the regulatory difficulties states face in reconciling science, health, culture and trade in the era of globalization. Technological change creates new products faster that our collective ability to assess their implications; new forms of transportation and expanding markets allow these products, and new pathogens, to move rapidly around the world because of the ever increasing exchanges of goods and services in the global economy. Information can be disseminated rapidly, but consensual knowledge does not keep up. Some regulatory decisions are effectively taken within gigantic multinational firms, or within such diverse international organizations as the FAO, the WHO, the ISO or the WTO, and other decisions are effectively pre-empted by civil society organizations, some of which are big multinationals in their own right.

> International cooperation is affected by questions of whether UNCED or the WTO should take precedence — is trade more important than the environment, or health? Other linked issues include intellectual property rights in seeds, and the approvals process for new pharmaceuticals. Labelling requirements for food have implications for eco-labelling schemes (e.g., certifying that wood came from sustainably managed forests) while agreements on food inspection may set precedents for general principles under the Technical Barriers to Trade agreement (e.g., mutual recognition of testing for conformity to product standards)....

> Countries have very different traditions and infrastructure for application of food regulatory regimes. Some countries use a market-based approach while others have had an interventionist approach to food inspection and consumer protection generally. Now increased trade flows are exposing the problems with purely national approaches to regulation.

## OVERVIEW OF THE PREDOMINANT THEMES IN THE DISCUSSIONS

The predominant themes in the seven discussion sessions as a whole may be represented as follows:

- science-based regulation *versus* consumer sovereignty
- risk assessment *versus* other factors in decision-making

These two representations are thematically similar, although they do not have exactly the same connotations, as will be seen in the remarks below.

The tension inherent in the first of the two primarily plays out largely in the marketplace of public opinion, and the second, largely in the risk management decision-making that is made both at national government levels and in the international organizations concerned with both food safety and with trade rules.

## SCIENCE-BASED REGULATION *VERSUS* CONSUMER SOVEREIGNTY

Science-based regulation refers to the underlying framework for assuring food safety, agreed to between governments and industry in western nations over the last 50 years and increasingly used as the basis of all international trade in food products. "Science" here refers to the processes of hazard identification and hazard characterization: in plain language, it is scientific research which will tell us, definitively, *what* potentially harmful things (hazards) we should primarily worry about in the matter of food safety, and what we can worry about less, or at all. The concept of science-based regulation, therefore, contains the notion that formal regulatory schemes, through which governments accountable to the public can assure their citizens that they are taking the right steps to protect health and the environment, will be based on the findings of science. This orientation also refers implicitly to "peer-reviewed" science, that is, scientific research findings that have been published in respected journals and affirmed on occasion in further peer reviews, such as those conducted by expert panels appointed by national academies.

In the last 20 years or so governments have routinely referred to "scientific findings" in their communications to the public on food safety, explicitly encouraging the public to rely on this standard when forming their own judgements about the safety of the food supply. However, experience has shown that consumers do not necessarily find these communications to be wholly persuasive. The conference presentation by Patricia Mann (former Vice-President International, J Walter Thompson), reporting on a major recent study on public perceptions in

Great Britain carried out by the new UK Food Standards Agency, pointed to widespread unease among the public about food safety in general, and in particular a huge increase in public concern about genetically modified (GM) ingredients in foods in the 1998–99 period. All major food retailing chains in the UK have removed foods with GM ingredients from their shelves, and have been assiduously tracing back product lines in their supply chains to ensure that they can advertise their wares as "GM-free." This, despite over ten years of consistent messages from industry and governments to the public to the effect that GM crops for food products have been carefully assessed as to safety, using science-based approaches, and have passed all of the relevant tests. These recent experiences have, in the words of Patricia Mann, clearly shown that factors other than science are important to consumers.

What factors? In the opening presentation Charles Cockbill, Chairman of the European Food Law Association of the UK, called attention to a basic truth: consumers have an approach to food and food safety very different from that which they have with respect to all other consumer products. Why this is so is not exactly clear, he added. However, in response to this truth, in many countries government food policies have switched from what might be called a "production orientation" to a "consumer orientation." This is reflected, for example, in governments taking steps to transfer responsibility for oversight of food safety from agriculture ministries, where they have been based for up to a hundred years, to health ministries. Some governments have taken the further step of transferring those responsibilities to stand-alone agencies which are to have a very high degree of independence from traditional "line" departments.

It is as yet unclear whether these steps will restore public confidence. Among other things, this is because science-based evidence does not seem to resonate well with many consumers, at least so far, because it does not seem to necessarily respond well to the diffuse and half-articulated consumer fears, especially when there are significant uncertainties in the scientific assessment. Messages from anti-government activists, for example, seem to be able to find a more sympathetic hearing, among at least some consumers. Lorne Hepworth, President of the Crop Protection Institute of Canada, reinforced this diagnosis, remarking generally that the "new global activism," which has been especially vocal in the series of anti-globalization demonstrations (Seattle to Genoa), is simply not well understood, either by governments or industry, in terms of its origins or potential appeal to a wider range of citizens. He also conceded that individuals often get mixed messages from industry, for example, when a large multinational corporation with different business units sells GM seeds to farmers through its agricultural unit, and at the same time bans GM ingredients in the food products sold through another one of its affiliates.

A pointed illustration about the limited impact of science-based regulation was provided in the talk by Rob McNabb, Assistant Manager of the Canadian Cattlemen's Association, who reviewed the long history of the battle between Europe and Canada over hormones in beef, which began in the early 1980s and is still not concluded. (McNabb reported the estimated economic losses to the Canadian beef industry at $75 million in 1999 and $1billion for the entire period.) In the 1990s there was a transition to science-based dispute resolution mechanisms for international disputes, including those over food, especially under the World Trade Organization (WTO). The decisions taken in these contexts have supported the Canadian position in general, and in the latest round, a deadline of compliance for the EU was set for May 1999; however, the European Union (EU) has taken no steps so far to comply with this decision.[1]

Discussion at the colloquium provided interesting insights on the reality of, and some of the reasons for, the existence of the tension between science-based regulation and consumer sovereignty. Questions were raised about whether consumers *should* at least have a choice in the marketplace, when something as "sensitive" as hormones is at issue. Giving consumers a choice would require, for example, the labelling of North American beef, with wording something like "this meat contains hormones administered as growth regulators," which the industry has not wished to do — precisely because, *on scientific grounds*, there is no reason to do so, because hormone residues in North American beef are no higher than in beef raised without such treatments. However, some individuals at the colloquium maintained that the "consumer's right to know" could override a rationale based on this scientific reasoning, without necessarily offending established trade rules.[2] Perhaps there would be no market for hormone-administered beef in Europe, if full disclosure were to be required. To say that such a consumer-driven rationale must necessarily be overridden by the science-based rationale would be to admit to a "scientization of politics," in the words of one discussant, a development that would be unacceptable.

A general consensus emerged from these discussions, to the effect that all risk management decisions by governments do occur, and will continue to occur, in a broad "political" context. This reinforced the theme in the earlier presentation by George Khachatourians, University of Saskatchewan, who posited that scientific work is increasingly thrown into an arena where the non-scientific dimensions of decision-making have equal or greater weight. Another presentation, by Spencer Henson of Reading University, gave some guidance as to what those other dimensions are, when he remarked that a transparent process of decision-making is perhaps the single most important ingredient of public confidence in any regulatory system. Thus, in the end, governments, no matter what regulatory structures (science-based or otherwise) they have established, will respond to the concerns of their citizens as expressed in their role as consumers of, for example, food products.

At present the consumer-sovereignty thrust is shown best in the demand for labelling of foods containing GM ingredients. Anne MacKenzie, of the Canadian Food Inspection Agency, speaking on "labelling of foods derived through bio-technology," reviewed the recent labelling initiatives in Switzerland, Japan, Brazil, Australia/New Zealand, and other countries. Demands for increasingly comprehensive labelling of food products, in many different areas, was the focus of another presentation, by Catherine Humphries of the Co-operative Wholesale Society. Whereas this demand has been strongly resisted by both industry and governments in North America, it is by now taken for granted in Europe that such labelling is appropriate. Some speculated that a "trade war" could emerge over the issue of labelling of products containing GM ingredients.

Finally, it seemed clear to some participants that one source of this tension between science-based regulation and consumer sovereignty was to be found in the fact that, increasingly, primary manufacturers of food system inputs (seeds, etc.) have strong relations with primary producers (farmers), but that these relationships stop at the farm gate. This is especially true with GM crops. The processors and retailers of food products, on the other hand, clearly do not want to have anything to do with the emergent consumer issues about GM foods. They have moved quickly to forbid the use of GM crops, such as Bt corn and Bt potatoes, in their consumer products, or to have products containing them sold in their stores. However, Douglas Powell of the University of Guelph had a different story to tell. Powell conceded that a "stigma" (negative connotation) could easily be attached to food products as a result of consumer worries, but argued that consumers will respond favourably to those who provide detailed, balanced, and clearly-communicated information to consumers about different types of food technologies. He reported on an experiment undertaken in the Guelph area, in which consumers were offered two different types of corn, one of which was genetically-modified, the other produced with conventional pesticides; consumers chose the former by a two-to-one margin.

## Conclusion

A series of questions, emerging from the colloquium discussions on these issues, could be posed for further reflection:

- Is there a risk of "reifying" science in what may be the excessive dependence of regulators on science-based decision-making?

- Are different *constructions* of science legitimate? (The "North American" approach identifies science with certainty, whereas others might emphasize that science is always provisional.)

- Do we expect too much of science in the North American approach?

- Is the emergence of the precautionary principle, as a new element in the decision-making mix,

    - "anti-science" or a different way of using science in decision-making?

    - an alternative to established risk management decision-making, or the expression of different underlying *social values* in different societies or regions?

    - as expressed in the Bio-Safety Protocol, for example, a decisive new element that will have real impacts on trade-related disputes involving approaches to risks?

## RISK ASSESSMENT *VERSUS* OTHER FACTORS IN DECISION-MAKING

Risk analysis is the general name for the process of risk-based decision-making now widely used for food safety oversight. The *Codex Alimentarius* (Codex) defines risk analysis as: "A process consisting of three components: risk assessment, risk management and risk communication:

1. *Risk Assessment*: A scientifically based process consisting of the following steps: (i) hazard identification; (ii) hazard characterization; (iii) exposure assessment; and (iv) risk characterization.

2. *Risk Management*: The process, distinct from risk assessment of weighing policy alternatives, in consultation with all interested parties, considering risk assessment and other factors relevant for the health protection of consumers and for the promotion of fair trade practices, and, if needed, selecting appropriate prevention and control options.

3. *Risk Communication*: The interactive exchange of information and opinions throughout the risk analysis process concerning hazards and risks, risk-related factors and risk perceptions, among risk assessors, risk managers, consumers, industry, the academic community and other interested parties, including the explanation of risk assessment findings and the basis of risk management decisions" (WTO 2000, p. 45).

The notion that there are "other relevant factors" is, as shown above, included in the Codex definitions. During the discussions, reference was made to the following (incomplete) list of such factors: environmental risks (as opposed to human health risks) such as biodiversity and use of pesticides and other chemicals; "national security" in food supply; the activities of global multinational corporations in food production and distribution; special circumstances of less-developed countries; the place of the farmer in society; farm subsidies; trade principles; nutrition and healthy diets; animal welfare; and the international transmission of plant and animal diseases.

A number of intervenors in the discussions on the tension between risk assessment and other factors pointed to one core issue: How is it possible, in a policy context, to "synthesize" the risk-based approach with these other, very different types of factors? Although by its very nature this is not a question that admits of an easy answer, or any answer at all, an interesting observation was made during these discussions: it sometimes seems that recently, at least in Europe, the "precautionary principle" or "precautionary approach" is serving as a surrogate or place-holder for an entire set of "other factors." The reason for this might be that the risk-based approach by definition can handle only well-characterized hazards, ideally ones that are suited to quantitative representation, and many items on the list of other factors do not fit this mold.[3] In this context, invoking the precautionary principle has the effect of saying, "slow down while we consider other factors." (This observation does not imply that only "well-characterized hazards" are appropriate entrants to the decision arena, quite the contrary.)

J.M. Scudamore, from the UK's Ministry of Agriculture, Fisheries and Food, showed how a risk management approach was being used in the design and mandate for the new Food Standards Agency (FSA), which has been given sole responsibility for food safety. The new agency, which is independent of line ministries, has the mandate to restore public confidence in food and, among other things, reflects the separation of food safety from promotion of agriculture. A similar, but not identical, design is being used for the proposed new European Food Authority (EFA), which like the FSA will rely upon independent science and will conduct an intensive dialogue with consumers on food issues. The EFA will have risk assessment and communication within its mandate, but risk management will remain the responsibility of the European Commission, a "politically accountable" body.

Neville Craddock, the Group Regulatory Affairs Manager of Nestlé UK, reinforced the view that a specific set of values lay behind the new European initiatives and agencies — namely, independence of science, transparency and openness in decision-making, and full public dialogue. He also seconded the importance of the precautionary principle, which had first been introduced into the discussion in a food safety context in the remarks by Spencer Henson, who had referred to the decision by the European Union's Court of Justice in upholding the ban on British beef due to concerns over bovine spongiform encephalopathy (BSE). The issue of BSE returned many times during these discussions, as something that has had a defining impact on the attitudes of Europeans toward food safety. As Craddock remarked, "assessments based on science" and "acceptable risk" are not now, and can never be in the post-BSE era, as straightforward or unproblematic as some would like them to be.[4]

Peter Phillips, University of Saskatchewan, supported these notions in a more systematic way, by pointing out that there appears to be emerging a multi-layered context for food safety issues, with a large number of influential players, no one of which is in a position to dominate the playing field. There are three main layers: a science-based one, such as IPPC (International Plant Protection Convention), OIE (Office international des épizooties), and Codex; a trade-based one (using risk-based approaches), principally WTO; and another set, with inherently broad mandates (OECD, Biodiversity Convention and BioSafety Protocol, and regional groups). In addition, the private sector is taking its own initiatives, on safety (use of HACCP and ISO) and through such acts as segregating GM and non-GM food products. This "portfolio" of responses includes, generally, both science- and risk-based ones and "consensual" ones.

Beatrice Olivastri from Friends of the Earth had intervened a number of times during the discussions over two days, urging the participants not to frame food safety issues too narrowly within the confines of current regulatory practices, especially those in North America. Food issues, she maintained, should always be framed as part of an environmental philosophy, which she called "a covenant with nature," where specific matters such as biodiversity protection, sustainable agriculture, and the interests of small farmers are always on the table in trade-related discussions. Rob Faulkner, from the University of Kent, summed up a good deal of the thematic unity in this section when he referred to the accusation, levelled against Europeans by the United States, that the "so-called" precautionary principle was simply an excuse for the Europeans' disregard of "appropriate" risk management decision-making. Falkner observed that the current opposition between Europe and North America on food issues reflects the fact that, in the EU, a multidimensional stakeholder-based political culture stands in sharp contrast to the top-down, risk-based, joint corporate/government consensus operative in North America. In this context, the various "takes" on the precautionary principle have *strategic* meaning in policy circles.

## Conclusions

There is an ongoing, structural problem in western countries about how to integrate science-based regulatory processes with public perceptions and consumer sovereignty. The increasing attention to labelling of GM foods and ingredients is one of the best illustrations of the gap between the North American risk managers, who have never accepted the rationale for any labelling, and most of the rest of the world, which is now trying to figure out how to do labelling in a way that actually assists consumers in expressing their preferences.

The proposed European Food Authority appears to be heading in the direction of separating responsibilities for *risk assessment* and *risk communication* from those for *risk management.* Risk assessment and risk communication — independent, credible science and assessment protocols and independent public dialogue resources for engaging citizens and raising the level of understanding — should be divided from risk management — incorporating risk assessment and public confidence within the context of other factors.

## GENERAL CONCLUSIONS

Our inherited regulatory structures do not permit us to achieve a unified perspective on the overall environmental and other consequences of different technologies in agriculture, because they evaluate risk-benefit trade-offs only within each technology, and not across different technologies. (Legal and other constraints are relevant here.) This is a good reason for also looking to independent assessment bodies for comparative risk-benefit assessments.

### NOTES

1. A detailed discussion of the European Community-hormones case under the WTO can be found in WTO (2000, pp. 26-29).
2. There is, in fact, support for this position in the decision of the WTO Appellate Body in the EC-hormones case. The panel that first looked at the matter, having found that the EC actions constitute the imposition of different levels of health protection, were required to give an opinion as to whether these differences were "arbitrary or unjustifiable," and they found that they indeed were so. However: "The Appellate Body disagreed. It stated that there was 'a fundamental distinction between added hormones (whether natural or synthetic) and naturally-occurred hormones in meat and other foods.' It therefore reversed the panel's finding on this first comparison" (WTO 2000, p. 37).
3. It is also the case that, under science-based regulatory regimes, applicants for approvals of products will appear before the regulators with a completed risk assessment (based

on known hazards) already in hand, asking for a quick decision to be made. It is only later, in the risk management phase, that the "other factors" are likely to be raised, which often, in the applicants' eyes, have the effect of "delaying" a decision.

4.   Perhaps it is too early to refer to the "post-BSE era" (Bremner 2000, p. A7).

## REFERENCES

Bremner, C. 2000. "France Falls Victim to Mad Cow Panic," *The Ottawa Citizen*, 10 November.

World Trade Organization (WTO). Committee on Sanitary and Phytosanitary Measures. 2000. "Summary Report on the SPS Risk Analysis Workshop, 19-20 June 2000," 3 November (G/SPS/GEN/209).

# Contributors

*Katija Blaine,* Research Assistant, Food Safety Network, University of Guelph

*Charles Cockbill,* President, European Food Law Association of the United Kingdom; Vice-President of the Executive Committee of the European Food Law Association

*Neville Craddock,* Group Regulatory and Environmental Affairs Manager, Nestlé UK Ltd

*Robert Falkner,* Lecturer in International Relations, University of Essex; Associate Fellow, Energy and Environment Programme, Royal Institute of International Affairs (Chatham House)

*Spencer Henson,* Associate Professor, Departments of Agricultural Economics and Business and Consumer Studies, University of Guelph

*Lorne H. Hepworth,* President, CropLife Canada (formerly the Crop Protection Institute)

*Catherine Humphries,* Chief Scientific Adviser, Co-operative Wholesale Society Ltd

*George G. Khachatourians,* Professor and Head, Department of Applied Microbiology and Food Sciences, College of Agriculture, University of Saskatchewan

*William Leiss,* Professor, School of Policy Studies, Queen's University; NSERC/SSHRC Research Chair in Risk Communication and Public Policy, University of Calgary

*Amber Leudtke,* Research Assistant, Food Safety Network, University of Guelph

*Anne A. MacKenzie,* Associate Vice-President of Science Evaluation, Canadian Food Inspection Agency; Chair of the *Codex Alimentarius* Commission's Committee on Food Labelling

*Patricia Mann*, OBE, former Vice-President International of J Walter Thompson

*Shane Morris,* Research Assistant, Food Safety Network, University of Guelph

*Peter W.B. Phillips*, Professor and NSERC/SSHRC Chair in Managing Knowledge-based Agri-food Development, University of Saskatchewan

*Douglas A. Powell,* Assistant Professor, Department of Plant Agriculture; Director, Food Safety Network, University of Guelph

*Jeff Wilson*, fruit and vegetable grower, Orton, Ontario

*Robert Wolfe*, Associate Professor, School of Policy Studies, Queen's University